ALL ABOUT TRADITIONAL TEXTILE FABRICS FOR DIY SPINNING, WEAVING, AND DYEING

CLASSIC INFORMATION ON FIBERS
AND CLOTH WORK

BY **PAUL N. HASLUCK**

ORIGINALLY PUBLISHED IN 1906

LEGACY EDITION

HASLUCK'S TRADITIONAL SKILLS LIBRARY
BOOK 8

Doublebit Press
Eugene, OR

New content, introduction, and annotations
Copyright © 2020 by Doublebit Press. All rights reserved.

Doublebit Press is an imprint of Eagle Nest Press
www.doublebitpress.com | Eugene, OR, USA

Original content under the public domain. Originally published in 1906 by Paul N. Hasluck under the title Textile Fabrics And Their Preparation for Dyeing.

This title, along with other Doublebit Press books including the Hasluck's Traditional Skills Library, are available at a volume discount for youth groups, outdoors clubs, or reading groups.

Doublebit Press Legacy Edition ISBNs
Hardcover: 978-1-64389-085-2
Paperback: 978-1-64389-086-9

Disclaimer: Because of its age and historic context, this text could contain content on present-day inappropriate methods, activities, outdated medical information, unsafe chemical and mechanical processes, or culturally and racially insensitive content. Doublebit Press, or its employees, authors, and other affiliates, assume no liability for any actions performed by readers or any damages that might be related to information contained in this book. This text has been published for historical study and for personal literary enrichment toward the goal of preserving the American handcraft tradition, timeless trade skills, and traditional artisanal knowledge.

First Doublebit Press Legacy Edition Printing, 2020

Printed in the United States of America
when purchased at retail in the USA

INTRODUCTION
To The Doublebit Press Legacy Edition

The old experts of artisanal trades, country and homestead knowledge, and the woods and mountains taught timeless principles and skills for centuries. Through their timeless books, the old experts offered rich descriptions of how the world works and encouraged learning through personal experiences *by doing*. Over the last 125 years, manufacturing, farming, and construction have substantially changed. Of course, many things have gotten simpler as equipment and technology have improved. In addition, some activities of pre-digital times are now no longer in vogue, or are even outright considered inappropriate or illegal. However, despite many of the positive changes in manufacturing and crafting methods that have occurred over the years, *there are many other skills and much knowledge that have been forgotten.*

By publishing *The Hasluck Traditional Skills Library*, it is our goal at Doublebit Press to do what we can to preserve and share the works from forgotten teachers that form the cornerstone of the history of the American artisans and traditional crafts. Through remastered reprint editions of timeless classics, perhaps we can regain some of this lost knowledge for future generations.

This book is an important contribution traditional handcraft and country skills literature and has important historical and collector value toward preserving the American handcraft and outdoors tradition. The knowledge it holds is an invaluable reference for practicing skills and hand craft methods. Its chapters thoroughly discuss some of the essential building blocks of knowledge that are fundamental but may have been forgotten as equipment gets fancier and technology gets smarter. In short, this book was chosen for Legacy Edition printing because much of the basic skills and knowledge it contains has been forgotten or put to the wayside in trade for more modern conveniences and methods.

With technology playing a major role in everyday life, sometimes we need to take a step back in time to find those basic building blocks used for gaining mastery – the things that we have luckily not completely lost and has been recorded in books over the last two centuries. These skills aren't forgotten, they've just been shelved. *It's time to unshelve them once again and reclaim the lost knowledge of self-sufficiency.*

Based on this commitment to preserving our outdoors and handcraft artisanal heritage, we have taken great pride in publishing this book as a complete original work. We hope it is worthy of both study and collection by outdoors folk in the modern era of outdoors and traditional skills life.

Unlike many other photocopy reproductions of classic books that are common on the market, this Legacy Edition does not simply place poor photography of old texts on our pages and use error-prone optical scanning or computer-generated text. We want our work to speak for itself, and reflect the quality demanded by our customers who spend their hard-earned money. With this in mind, each Legacy Edition book that has been chosen for publication is carefully remastered from original print books, *with the Doublebit Legacy Edition printed and laid out in the exact way that it was presented at its original publication.* We provide a beautiful, memorable experience that is as true to the original text as best as possible, but with the aid of modern technology to make as beautiful a reading experience as possible for books that can be over a century old.

Because of its age and because it is presented in its original form, the book may contain misspellings, inking errors from print plates, and other printing blemishes that were common for the age. However, these are exactly the things that we feel give the book its character, which we preserved in this Legacy Edition. During digitization, we ensured that each illustration in the text was clean and sharp with the least amount of loss from being copied and digitized as possible. Full-page plate illustrations are presented as they were found, often including the extra blank page that was often behind a plate. For the covers, we use the original cover design to give the book its original feel. We are sure you'll appreciate the fine touches and attention to detail that your Legacy Edition has to offer.

For traditional handcrafters and classic artisanal enthusiasts who demand the best from their equipment, this Doublebit Press Legacy Edition reprint was made with you in mind. Both important and minor details have equally both been accounted for by our publishing staff, down to the cover, font, layout, and images. It is the goal of Doublebit Legacy Edition series to be worthy of collection in any outdoorsperson's library and that can be passed to future generations.

Every book selected to be in this series offers unique views and instruction on important skills, advice, tips, tidbits, anecdotes, stories, and experiences that will enrichen the repertoire of any person who enjoys escaping a bit from today's modern technology-based, cookie-cutter, and highly industrialized skills. Instead, folks seeking to make things with their hands like the old days may find great value from these resurrected instructional manuals from the past. These books were not simply written to be shelved in a library – they contain our history and forgotten methods to make things with real character and energy with a *human* component.

Therefore, to learn the most basic building blocks of a craft leads to mastery of all its aspects. We hope this book helps you along this path with its rich descriptions and illustrations!

About Hasluck's Traditional Skills Library

Paul N. Hasluck was a prominent author on artisan skills and traditional handcrafts toward the end of the 19th Century. He was the editor of the magazine *Work*, which was a popular handcraft, shop skills, and artisanal craft magazine of the day. His broad expertise in making things with your hands led him to write or edit over 30 volumes on specific handcrafts, arts, and mechanics, with each manual containing invaluable information related to each craft.

Hasluck had a great eye for collecting the info that beginners and experts alike needed to perfect their craft. His volumes were loaded with helpful diagrams, tables, and illustrations that are useful even by today's digital standards. In short, Hasluck's instructional manuals were the *go-to instructional library* if someone wanted to learn a particular skill. Used by the U.S. military, the Boy and Girl Scouts, and countless folks at farms, public libraries, and homes across the world, Hasluck's instructional manuals were the perfect "handy book" for learning.

This Doublebit Press Legacy Edition republishes this tradition of handcrafted quality and artisanal work. We hope that this deluxe printed edition of this work will help you gain mastery in your craft, as it is presented in the exact form that it was originally published. Even today, the knowledge contained within its pages are timeless and have much to teach!

Finally, as art, Hasluck's manuals contain beautiful illustrations and line art that are a sign of simpler, yet authentic times when quality mattered and craftsmanship was king. This collectible volume makes a great addition to the bookshelf of any handcrafter, maker, artisan, farmer, homesteader, or outdoors enthusiast!

PREFACE.

TEXTILE FABRICS AND THEIR PREPARATION FOR DYEING contains, in a form convenient for everyday use, a comprehensive treatise on the subject. The contents of this manual are based on the highly esteemed book written by the late Dr. J. J. Hummel, F.C.S., Professor and Director of the Dyeing Department of the Yorkshire College, Leeds.

Without omitting any essential part of the original work the matter has been revised and brought up to date by Mr. A. R. Foster, Consulting Textile Expert, City and Guilds Honours Medallist. Needless to say many changes have taken place since the previous edition was published, and whilst the new processes and appliances have been incorporated, the older methods which are still in vogue in less progressive works, have been retained and revised. In this manner the manual has been made valuable, not only to the student but all employed in bleaching, finishing, and dyeing works.

Readers who may desire additional information respecting special details of the matters dealt with in this book, or instructions on any kindred subjects, should address a question to the Editor of WORK, La Belle Sauvage, E.C., so that it may be answered in the columns of that journal.

P. N. HASLUCK.

La Belle Sauvage, London,
  November, 1906.

CONTENTS.

CHAPTER	PAGE
I.—Cotton	9
II.—Flax, Jute, and China Grass	20
III.—Wool	29
IV.—Silk	47
V.—Cotton Bleaching	71
VI.—Linen Bleaching	88
VII.—Mercerising	93
VIII.—Wool Scouring and Bleaching	96
IX.—Scouring and Bleaching Silk	119
X.—Water	124
XI.—About Dyeing	145
Index	15

LIST OF ILLUSTRATIONS.

FIG.		PAGE
1.	Cotton Plant	9
2.	Appearance of Cotton under the Microscope	10
3.	Transverse Sections of Cotton Fibre	11
4.	Transverse Sections of Unripe Cotton Fibre	11
5.	Transverse Sections of Cotton Fibre after Treatment with Caustic Soda	17
6.	Flax Fibre under the Microscope	24
7.	Microscopical Appearance of Wool Fibre	30
8.	Cells of Wool Fibre under the Microscope	32
9.	Cross Section of Typical Wool Fibres	33
10.	Silk Glands of the Silkworm	48
11.	Section of Silk-bag	49
12.	Microscopic Appearance of Raw Silk Fibre	49
13.	Silk Cocoon	50
14.	Microscopic Appearance of Tussur Silk Fibre	54
15.	Stringing Machine for Silk	56
16.	Details of Silk-stringing Machine	57
17.	Silk-lustring Machine	59
18.	Conditioning Apparatus	60
19.	Section of Conditioning Chamber	63
20.	Apparatus for Chemicking, Souring, and Washing	72
21.	Plate Singeing Machine, by Mather and Platt	
22.	Gas Singeing Machine, by Mather and Platt	
23.	Longitudinal Section of the Mather Patent Kier	79
24.	Section of Injector Kier	82
25.	Five-bar Expander, by Mycock	85
26.	Wool-steeping Tank—Elevation	101
27.	Wool-steeping Tank—Plan	101

FIG.		PAGE
28.—Section of Furnace for Making Yolk-ash		102
29.—McNaught's Wool-scouring Machine—Elevation		104
30.—McNaught's Wool-scouring Machine—Cross Section		105
31.—Yarn-stretching Machine		107
32.—Woollen Yarn-scouring Machine		109
33.—Continuous Woollen Yarn-scouring Machine		110
34.—Woollen Cloth-scouring Machine		110
35.—Section of Machine shown in Fig. 34		111
36.—Woollen Cloth Open-width Scouring Machine		112
37.—Treble Crabbing Machine		114
38.—Sulphur Stove for Woollen Cloth Bleaching		116
39.—Porter-Clark's Apparatus for Softening Water—Plan		134
40.—Porter-Clark's Apparatus for Softening Water—Elevation		135
41.—Gaillet and Huet's Apparatus for Softening Water		138
42.—Gaillet and Huet's Precipitating Tank		139
43.—Purification Works for Waste Dye-liquors		142

CHAPTER I.

COTTON.

The Cotton Plant.—Cotton is the white, downy, fibrous substance which envelopes the seeds of various species of the cotton-plant, *Gossypium*, belonging to the natural

Fig. 1.—Cotton Plant.

order *Malvaceæ*. The seeds, to which the cotton fibres are attached, are enclosed in a 3- to 5-valved capsule, which bursts when ripe; the cotton is then collected and spread out to dry. The seeds are afterwards separated by the mechanical operation termed "ginning," and the raw cotton thus obtained is sent to the spinner. The cotton-plant (Fig. 1) is cultivated with success only in warm climates. There are numerous varieties, of which the following are the principal:—

(1) *Gossypium barbadense.*—An herbaceous plant, bearing a yellow flower, and attaining a height of 4-5 m. (13-16 ft.). A variety of this species yields the Sea Island cotton, much prized on account of the great strength, length, and lustre of its fibres. It is grown in the North American States of South Carolina, Georgia, and Florida, and on the neighbouring islands of the West Indies.

(2) *Gossypium hirsutum.*—A hairy, herbaceous plant, about 2 m. high, with pale yellow or almost white flowers. It is grown in the States of Alabama, Louisiana, Texas, and Mississippi.

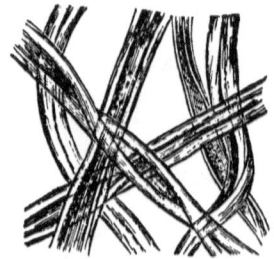

Fig. 2.—Appearance of Cotton under the Microscope.

(3) *Gossypium herbaceum.*—A small herbaceous plant, 1 m. high, and bearing yellow flowers. Varieties of this species are grown in India, China, Egypt, and America. The Madras, Surat, and short-stapled Egyptian cotton, also some American cottons, are obtained from this species.

(4) *Gossypium peruvianum.*—This species, a native of South America, grows to a height of 3-5 m. (10-16 ft.), and bears a yellow flower. It yields the long-stapled and much-esteemed Peruvian and Brazilian cottons.

(5) *Gossypium religiosum.*—This is a low annual shrub, about 1 m. high, and bearing a yellow flower. It is grown in China and India, and yields the so-called Nankin cotton, remarkable for its tawny colour.

(6) *Gossypium arboreum.*—This is a perennial tree, growing to a height of 6-7 m. (20-23 ft.), and bearing reddish-purple flowers. It is a native of India, and produces a good quality of cotton.

Physical Structure of Cotton.—If raw cotton is examined under the microscope, it is seen to consist of minute fibres. Their general appearance is that of spirally-twisted bands, having thickened borders and irregular markings on the surface (Fig. 2). In the better qualities of cotton, such as Sea Island, the spiral character is less prominent. Transverse sections of the fibres show them to be flattened tubes, having comparatively thick walls and a small central opening (Fig. 3).

A single cotton fibre is an elongated, tapering, and

Fig. 3.—Transverse Sections of Cotton Fibre.

collapsed plant cell, the thin end of which is closed, and the other (by which it was attached to the seed) irregularly torn. Sometimes broad ribbon-like fibres may be noticed, which are remarkably transparent, and possess irregular folds. Their transverse section exhibits no central opening (Fig. 4). They are unripe fibres, in which no separation of the thin cell walls has yet taken place. They refuse to be dyed like ordinary ripe fibres, and appear occasionally as white specks in indigo- and madder-dyed calicoes; hence the name of dead cotton has been given

Fig. 4.—Transverse Sections of Unripe Cotton Fibre.

to them. In half-ripe cotton fibres the cell walls are still so closely pressed together that the ultimate central canal is indicated in a transverse section only by a fine line. When steeped in water, however, such fibres gradually swell up and form hollow tubes. Cotton fibres vary in length 2·5-6 cm. (1-2½ in.), and in breadth 0·017-0·05 mm. (·0007-·002 in.).

The spiral character of the fibre makes it possible to spin exceeding fine yarn, and also accounts for the elastic character of calico as compared with linen, the fibres of which are stiff and straight. The microscopic appearance of cotton serves to distinguish it from other vegetable and animal fibres.

Chemical Composition of Cotton.—The substance of the cotton fibre is called Cellulose. This is almost universal in vegetable cells, forming the so-called ligneous matter or woody fibre of plants, but whereas in woody fibre the cellulose is encrusted with a large proportion of foreign matter —such as dried-up sap, resin, etc.—in the cotton fibre it is in a tolerably pure condition. The impurities present amount to about 5 %, this being the loss sustained by raw cotton when submitted to the process of bleaching, the main object, indeed, of which is the total removal of these impurities. The principal bleaching operation consists in boiling the cotton with a solution of sodium carbonate or hydrate. From the dark brown solution thus obtained, acids throw down a voluminous light brown precipitate, which, when washed and dried, amounts only to about 0·5 % of the weight of cotton employed. This precipitate is found to consist of the following organic substances: Pectic acid, brown colouring matter, cotton wax, fatty acids (margaric acid), and albuminous matter. Pectic acid exists in the largest proportion, and it is not improbable that the 4·5 % loss by bleaching still unaccounted for, represents certain pectic matters, modified and rendered soluble by the action of alkalis, but not precipitated by acids.

In addition to the above-mentioned impurities of the cell wall, the raw cotton fibre seems to be covered with an exceedingly delicate membrane, or cuticle, which is not cellulose. If cotton, when under microscopical observation, be moistened with an ammoniacal solution of cupric hydrate, the fibre swells up under its influence, whereas the cuticle is unaffected and shows itself as band-like strictures or rings of various breadths. If a drop of sulphuric acid be then added, the cellulose separates out as a gelatinous mass, which, on adding a drop of iodine solution, becomes coloured blue, whereas the cuticle is coloured yellow. By moving the cover-glass aside a little, the cuticle rings are seen to be in the form of tubes, possessing apparently a spiral structure. Some observers state that during the bleaching process this cuticle is removed, while others say this is not the case. The standard of moisture in raw cotton is 8·5 %, so that, reckoning the 5 % impurities already alluded to, one may consider that raw cotton contains 86·5 % of pure dry cellulose.

When submitted to chemical analysis, cellulose is found to be composed of carbon, hydrogen, and oxygen, the formula assigned to it being $C_6H_{10}O_5$. It is closely allied in composition to starch, dextrin, and glucose, and is classed along with them as a carbo-hydrate. It is colourless, possesses neither taste nor smell, and has a density of about 1·5. If heated above 130° C. it becomes brown, and begins to decompose. In contact with air it burns without emitting any very strong odour, a fact which may sometimes serve to distinguish it from wool and silk. It is quite insoluble in the ordinary solvents, water, alcohol, ether, etc., but, as already indicated, it dissolves in an ammoniacal solution of cupric hydrate; from this it is precipitated by acids as a gelatinous mass, which, when washed with alcohol, forms an amorphous white powder.

Action of Mildew on Cotton.—Owing to its comparative freedom from impurity, cotton may be stored for a long period without undergoing any change, more especially if it is bleached and kept dry. When, however, it is contaminated with added foreign organic matter, such as starch, gum, etc. (in " finished " calicoes), and then exposed to a moist, warm atmosphere, it is very liable gradually to become tender or rotten. This is owing to the growth of vegetable organisms of a very low order, generally called " mildew." These fungi feed upon the starchy matters present, inducing their decomposition, and after some time the cotton fibres themselves are attacked. The simultaneous production of crenic, humic, ulmic, and other organic acids may possibly assist somewhat in the tendering process.

Action of Frost on Cotton.—It has been supposed by some that wet calico is tendered when it is frozen. Although the evidence on this point is conflicting, it is quite conceivable that the crystallisation might act injuriously in a mechanical way, and that the atmospheric ozone might also exercise some slight destructive influence. The popular notion probably arises from the fact that in their rigid state the cotton fibres are readily broken. A similar friable condition is obtained by excessive stiffening with starch or gum.

Action of Acids on Cotton.—Cold dilute mineral acids have little or no action, but if allowed to dry upon the cotton they gradually become sufficiently concentrated to corrode and tender the fibre. The physical structure of the fibre is not affected, but the chemical composition of the disintegrated fibre seems to be somewhat altered: it contains more oxygen and hydrogen. The same corrosive action soon takes place if cotton impregnated with such acids is heated. The process of "extracting" or "carbonising" woollen rags containing cotton (destroying and removing the cotton), by means of sulphuric or hydrochloric acid, is founded on this fact.

The action of strong acids varies considerably according to the nature, concentration, and temperature of the acid, as well as the duration of its contact with the fibre.

Very concentrated sulphuric acid causes cotton to swell up, and form a gelatinous mass, from which, on the addition of water, a starch-like substance termed Amyloid may be precipitated. A solution of iodine colours this amyloid blue. Vegetable parchment is paper (cellulose) superficially changed into amyloid by a short steeping in strong sulphuric acid 140° Tw. (Sp. Gr. 1·7), then washing and drying. An increased affinity for basic coal-tar colouring matters is said to be imparted to cotton by this treatment, even when the acid is diluted to 84° Tw. (Sp. Gr. 1·42), although its physical aspect then remains unchanged. Cotton completely disorganised by acid, and obtained as a fine powder, seems to contain one molecule of water more than ordinary cellulose, and the substance thus produced has been termed **Hydro-cellulose.**

If the concentrated sulphuric acid is allowed to act for a longer time, the cotton dissolves with the formation of a different substance of a gummy nature, called Dextrin ($C_6H_{10}O_5$); when the solution is diluted with water and boiled for some time, this dextrin is further changed into Glucose ($C_6H_{12}O_6$).

If cotton be heated with strong nitric acid it is entirely decomposed, producing oxalic acid and an oxidised cellulose soluble in alkalis. By the action of cold concentrated nitric acid, or, better still, a mixture of strong nitric and sulphuric acids, cellulose is changed into so-called Nitrocellulose. The physical structure of the cotton remains the same, although increased in weight by more than 5 %, but its chemical composition and properties are very much altered, certain elements of the nitric acid having replaced a greater or less proportion of the hydrogen of the cellulose. The most highly nitrated compound is Pyroxylin, or Gun-cotton ($C_{12}H_{14}(NO_2)_6O_{10}$), produced by the short action of a very concentrated mixture of acids; it is very explosive, and insoluble in alcohol and ether. The less nitrated product, obtained by the longer action of more dilute acids, forms the so-called Soluble Pyroxylin. Its solution in a mixture of ether and alcohol constitutes Collodion, which on evaporation leaves the pyroxylin as a thin, transparent, horny film, insoluble in water. It was long ago noticed by Kuhlmann that gun-cotton had an increased affinity for colouring matters, but no practical use has been made of the fact.

Strong hydrochloric and phosphoric acids behave towards cotton like sulphuric acid, but their action is less energetic. The ultimate product of the action of hydrochloric acid on cotton is the same as that given by sulphuric acid, but there is no intermediate formation of amyloid.

Solutions of tartaric, citric, and oxalic acids have no destructive action on cotton if it is simply steeped in the liquid; but if cotton saturated with a solution containing 2 % of any of the above acids is dried and heated for an hour to 100° C., it becomes slightly tendered. With 4 % solutions the destructive action is very decided at 100° C., and perceptible even at 80° C. Oxalic acid has the most injurious effect in this respect. If the acid solutions are thickened with gum or starch, and if steaming be substituted for a dry heat, the corrosive action in each case is less marked, so that in the ordinary practice of the calico-printer, who frequently uses steam-colours containing 4 % or more of the above acids, there is little to fear. Still, it is well to bear in mind that even organic acids cannot under all circumstances be applied to cotton with impunity.

Acetic acid may be considered as having no perceptible action on cotton.

Action of Alkalis on Cotton.—Weak solutions of caustic potash or soda, when used cold, have, under ordinary circumstances, no action on cotton, although long-continued and intermittent steeping and exposure to air tender the fibre. Cotton may even be boiled for several hours with weak caustic alkalis, if care be taken that it remains steeped below the surface of the solution during the whole operation, but otherwise it is very liable to become rotten, especially if the exposed portions are at the same time under the influence of steam. Such exposure is to be guarded against during certain of the operations in bleaching cotton fabrics. The tendering action is probably due to oxidation. It is worthy of note that the disorganised fibre (oxy-cellulose) possesses an increased attraction for basic coal-tar colouring matters.

In the case of raw cotton, the action of boiling with weak caustic alkalis is simply to remove those natural impurities already referred to, which cause it to be water-repellent, and therefore difficult to wet by mere steeping in cold water.

The action of strong solutions of caustic potash or soda is very remarkable. If a piece of calico is steeped for a few minutes in a solution of caustic soda, marking about 50° Tw. (Sp. Gr. 1·25), it assumes quite a gelatinous and translucent appearance; when taken out and washed free from alkali, it is found to have shrunk considerably, and become much closer in texture. If a single fibre of the calico thus treated be examined under the microscope, it is seen to have lost all its original characteristic appearance; it has no superficial markings, and is no longer flat and spirally twisted, but seems now to be thick, straight, and transparent. A transverse section shows it to be cylindrical, while the cell walls have considerably thickened, and the central opening is diminished to a mere point (Fig. 5). Many years ago a Lancashire calico-printer, John Mercer, discovered that calico treated in the above manner was not only stronger than before, but had also acquired an increased attraction for colouring matters. Hoping to apply the process with advantage preparatory to dyeing, he patented it. Cotton thus treated was at that time said to be Mercerised, but that term is now applied to cotton or cotton materials which are treated with soda lye under a tension which prevents the shrinkage above mentioned, or are afterwards stretched to counteract that shrinkage. When dyed in the indigo-vat, mercerised calico requires only one dip to produce as deep a shade of blue as can be obtained on ordinary calico only after five or six dips. Again, if a piece of ordinary and a piece of mercerised calico be dyed alizarin red, making all other conditions (time, temperature, quantity of Alizarin, etc.) the same in both cases, the mercerised cloth will be found to have a much fuller and richer colour than the other. Similar differences in depth of shade are noticed with other colours. The process, however, was never adopted in general practice as a means of saving dyestuffs, but in more recent years has been largely adopted for giving a brilliant lustre to the cotton.

Caustic ammonia in aqueous solution, whether strong or weak, has, under all circumstances, no action on cotton.

Fig. 5.—Transverse Sections of Cotton Fibre after Treatment with Caustic Soda.

Dry cotton is said to absorb 115 times its bulk of ammonia gas. Solutions of the carbonates of potash, soda or ammonia, silicate of soda, borax, and soap, have practically no action on cotton. There is a case on record, however, in which calico impregnated with silicate of soda, and shipped from England to South Africa, was found, after having been packed in bales for two years, to have become tender. Examination showed that the silicate of soda had decomposed with formation of silicic acid and carbonate of soda, and it was concluded that the tendering was due partly to the long-continued action of the carbonate of soda on the cotton, and partly to disruption of the fibres by the expansive force of the crystallisation of the carbonate of soda formed within them. The explanation is not altogether satisfactory, since it was found impossible to produce the same effects artificially. Probably it was a case of oxidation of the fibre. Whatever may have been the real cause, it is well to bear in mind that under exceptional conditions like those mentioned, even apparently harmless salts may tender the cotton fibre.

Action of Lime on Cotton.—Milk of lime, even at a boiling heat, has little or no action upon cotton so long as the latter is steeped below the surface of the liquid, but if it is at the same time exposed to the action of air or steam, it becomes much tendered by oxidation of the fibre. Such exposure must be avoided in cotton-bleaching.

Action of Chlorine and Hypochlorites on Cotton.—Cotton is quickly tendered if exposed to moist chlorine gas, especially in strong sunlight. The action may be due partly to the direct action of chlorine upon the fibre, one portion combining with and another replacing some of its hydrogen, partly to the destructive action of the hydrochloric acid thus produced, and partly to oxidation. Solutions of hypochlorites (bleaching powder, etc.) tender cotton more or less readily, according to the strength and temperature of the solutions and the duration of their action. Even a very weak solution of bleaching-powder will tender cotton if the latter be boiled with it; but when used cold, even if it be at the same time exposed to the air, the destructive action is inappreciable, and confined merely to bleaching the natural colouring matter of the cotton. If a piece of calico is moistened with a solution of bleaching-powder to 5° Tw. (Sp. Gr. 1·025), then exposed to the air for about an hour, and washed, it will be found to have acquired an attraction for basic coal-tar colouring matters similar to that possessed by the animal fibres. Cotton thus treated also decomposes directly, the normal salts of aluminium, iron, etc., attracting metallic oxide. Experiment has shown that this remarkable change is due to the action of the hypochlorous acid liberated by the carbonic acid of the air. The cotton thereby becomes chemically changed to what has been called by Witz, its discoverer, Oxy-cellulose.

Action of Metallic Salts on Cotton.—Under ordinary circumstances, solutions of neutral salts have no action on cotton; even those of acid salts have no appreciable effect if the cotton be merely steeped in them while cold; but if boiled with them the effect is similar to that of the free acids, though slightly less marked. If cotton is impregnated with solutions of the salts of the earths and heavy metals, then dried, and heated or steamed, the salts are readily decomposed; a basic salt is precipitated on the fibre, and the liberated acid affects the fibre according to the nature and strength of the salt solution employed.

The use of aluminium chloride, which was at one time recommended for the purpose of destroying the cotton in rags containing cotton and wool ("extracting"), also the application of the "topical" or "steam colours," and the "mordanting" process employed by the calico-printer, are all based upon the above facts.

Action of Colouring Matters on Cotton.—With few exceptions, colouring matters are not directly attracted from their solutions by the cotton fibre, hence it is not readily dyed, and special means of preparing it to receive the dyes have to be adopted in most cases (mordanting). The reason of this inert character of cotton is not yet satisfactorily explained; probably both its chemical and physical structure have an influence in the matter.

CHAPTER II.

FLAX, JUTE, AND CHINA GRASS.

The Flax Plant.—The term "flax" designates the flax or linen fibre and also the plant from which it is obtained. Linen fibre consists of the bast cells of certain species of the genus *Linum*, more particularly *Linum usitatissimum*, a plant belonging to the natural order *Linaceæ*. It is an herbaceous plant, having a thin, spindle-shaped root, a stem usually branched at the top, smooth lanceolate leaves, and bright blue flowers, and is cultivated in nearly all parts of Europe.

The time of sowing varies in different countries from February to April, consequently the time of harvest also varies, and may be from June to September.

If the object of the farmer is to obtain good fibre, and not seed for re-sowing, the plant is gathered before it is fully matured—namely, when the lower portion of the stem (about two-thirds of the whole) has become yellow, and the seed capsules are just changing from green to brown. At this stage the plants are carefully pulled up. If they are left in the ground till the plant is fully ripe and the whole stem is yellow, the fibre obtained will be more stiff and coarse.

The freshly-pulled flax is at once submitted to the process of "rippling," which has for its object the removal of the seed capsules. This operation is performed by hand, by drawing successive bundles of flax-straw through the upright prongs of large, fixed iron combs, or "ripples." If the pulled flax has been dried and stored, the removal of the seeds is usually effected by the seeding-machine, which consists essentially of a pair of iron rollers, between which the flax-straw is passed.

Retting.—The most important operation in separating the fibre is that of "retting," the object of which is to decompose and render soluble by fermentation, as well as to remove, certain adhesive substances which bind the bast fibres not only to each other, but also to the central woody portion of the stem, technically termed the "shrive," "shore," or "boon."

The various modes of retting may be classified as :—

(1) Cold-water retting, which may be carried out with running or with stagnant water.

(2) Dew retting.

(3) Warm-water retting.

Cold-water Retting.—The best system of retting in running water is said to be practised in the neighbourhood of Courtrai, in Belgium, where the water of the sluggish river Lys is available. The bundles of flax straw are packed vertically in large wooden crates lined with straw. Straw and boards are afterwards placed on the top, and the crate thus charged is anchored in the stream and weighted with stones, so that it is submerged a few inches below the surface. In a few days fermentation begins, and as it proceeds additional weight must be added from time to time, in order to prevent the rising of the crates through the evolution of gas. As a rule, after steeping for a short period, the flax is removed from the crates, and set up in hollow sheaves to dry; it is then repacked in the crates, and again steeped until the retting is complete. According to the temperature, quality of flax, etc., the duration of the steeping may be from 10-20 days. The end of the process must be accurately determined by occasionally examining the appearance of the stems, and applying certain tests. The flax bundles should feel soft, and the stems should be covered with a greenish. slime, easily removed by passing them between the finger and thumb; when bent over the forefinger the central woody portion should spring up readily from the fibrous sheath. If a portion of the fibre is separated from the stem and suddenly stretched, it should draw asunder with a soft, not a sharp, sound. When the retting is complete, the flax is carefully removed from the crates and set up in sheaves to dry.

Stagnant-water retting is the method usually adopted in Ireland and Russia. The flax is steeped in ponds, preferably situated near a river, and provided with suitable arrangements for admitting and running off the water.

This mode of retting is more expeditious than when running water is employed, because the organic matters retained in the water very materially assist the fermentation; there is, however, always a danger of "over-retting," that is, the fermentation may become too energetic, in which case the fibre itself is attacked and more or less weakened. This danger is minimised by occasionally changing the water during the steeping process. The quality of the water employed in retting is of considerable importance; pure soft water is the best, calcareous water being altogether unsuitable. The waste water, being strongly impregnated with decomposing organic matter, poisons the streams into which it may run, and destroys the fish; but it possesses considerable value as a liquid manure.

After retting in stagnant water, the flax is drained, then thinly spread on a field, where it is left for a week or more, and occasionally turned over. This process is termed "spreading" or "grassing." Its object is not merely to dry the flax, but to allow the joint action of dew, rain, air, and sunlight to complete finally the destruction and removal of the adhesive substances already alluded to. After a few days' exposure the stems begin to "bow," the fibrous sheath separates more or less from the woody centre, and the latter becomes friable.

Dew retting consists in spreading the flax on the field and exposing it to the action of the weather for 6-8 weeks, without any previous steeping. Damp weather is the most suitable for this method, since all fermentation ceases if the flax becomes dry. Dew retting is practised largely in Russia and in some parts of Germany.

Warm-water retting was a system recommended in 1847 by R. B. Schenck. It consists in steeping the closely-packed flax bundles in covered wooden vats, filled with water heated to 25°-35° C. By this means the fermentation is much accelerated, and the operation is completed in 2-3 days; the process seems, however, to have met with only limited success.

Chemical retting, the process recommended by R. Baur, may be mentioned. It consists in first squeezing the fresh or dried flax straw between rollers, and then steeping it in water till the latter ceases to be coloured yellow.

It is next drained and steeped for 1-2 days in dilute hydrochloric acid (3 kg. concentrated HCl per 100 kg. flax), until the bast fibres can be readily separated. The acid liquid is then run off, and the flax is well washed with slightly alkaline water, or such as contains a little chalk. A further treatment with dilute bleaching powder solution to dissolve away still adhering woody matter, and a final washing, complete the process. A well-retted flax is said to be thus obtained in the course of a few days only.

Chemistry of Retting.—Experiments by Kolb indicate that the adhesive matter which cements the flax fibres together is essentially a substance called pectose. During the retting process the fermentation decomposes this insoluble pectose, and transforms it into soluble pectine, and insoluble pectic acid. The former is washed away, the latter remains attached to the fibre.

Breaking.—The next operation is to remove the woody centre from the retted and dried flax, after which the fibres must be separated from each other. It is rather beyond the scope of this manual to give more than a general account of the nature of the various mechanical operations for effecting this. They comprise "breaking," "scutching," and "hackling."

The first operation aims at breaking up the brittle woody centre of the flax into small pieces, by threshing it with an indented wooden mallet, or by crimping it with a many-bladed "braque." The operation is now extensively done by machinery, the flax being passed through a series of fluted rollers.

Scutching.—In this process handfuls of the flax are beaten with a broad wooden scutching-blade; the particles of woody matter adhering to the fibres are thus detached; and the bast is partially separated into its constitutent fibres. Scutching is also performed by machinery. The waste fibre obtained is called "scutching tow," or "codilla."

Hackling.—The subsequent hackling, or heckling, has for its object a still further separation of the fibres into their finest filaments, by combing. When done by hand, a bundle of flax is drawn, first one end and then the other, through a succession of fixed upright iron combs or "hackles" of different degrees of fineness, beginning with the coarsest. When machinery is used, the flax is held against hackles fixed on moving belts or bars or on the circumference of revolving cylinders. The product of the operation is twofold, namely, "line" and "tow"; the former consist of the long and more valuable fibres, the latter of those which are short and more or less tangled.

Flax-line.—The appearance of flax-line is that of long, fine, soft, lustrous fibres, varying in colour from the yellowish-buff of the Belgian product to the dark greenish-grey of Russian flax. This difference in colour is chiefly owing to the system of retting adopted. Flax retted in running water has a more or less pale yellowish-buff

Fig. 6.—Flax Fibre under the Microscope.

colour, while that retted in stagnant water possesses a greyish colour, probably because of the presence of the decomposing organic matter in the water.

Physical Structure and Properties of Flax.—Examined under the microscope, a single flax fibre appears as a long, straight, transparent tube, often striated longitudinally (Fig. 6); it possesses thick walls, and an excessively minute central canal. At irregular intervals it is slightly distended, and at these points faint transverse markings may be detected. When examined with high powers, they seem to consist of a succession of very minute fissures, and, according to Vetillart, are simply breaks, or wrinkles, produced by bending of the fibre, and not cell divisions, or nodes, as frequently stated. Fibres which have been **vigorously** rubbed between the fingers, or have been subjected to the lengthened disintegrating action of alkalis, exhibit well-marked longitudinal fissures, and the broken end of a well-worn fibre presents the aspect of a bundle of fibrils. These appearances evidently indicate that the cell wall of the linen fibre possesses a fibrous structure.

The average length of a single fibre is 25-30 mm. (1-1·10 in.), and the average breadth 0·020-0·025 mm. (0·008-0·0082 in.). In transverse section, the linen fibre shows a more or less rounded polygonal contour.

The chief physical characteristics of the linen fibre, when freed from all encrusting material, are its snowy whiteness, silky lustre, and great tenacity. This last feature is no doubt owing to its fibrous texture as well as to the thickness of the cell walls. Its straight, even, prismatic, and transparent character accounts largely for the lustre.

Linen is hygrometric to about the same degree as cotton, and contains, when air-dry, 12 % of moisture. As it is a much better conductor of heat, it feels colder than cotton, and is also less pliant and less elastic.

Chemical Composition of Flax.—Treated with sulphuric acid and iodine solution, the thick cell wall is coloured blue, while the secondary deposits, immediately enclosing the central canal, acquire a yellow colour. The linen fibre consists, therefore, essentially of cellulose, but in its raw unbleached state it is mixed with about 15-30 % of foreign substances, chief among which is pectic acid. Fatty matter, to the extent of about 5 %, colouring matter, and other substances not investigated, are also present.

Action of Chemicals on Flax.—Being cellulose, the action of various chemical agents on pure linen fibre is much the same as on cotton, but, generally speaking, linen is more susceptible to disintegration, especially under the influence of caustic alkalis, calcium hydrate, and strong oxidising agents, such as chlorine, hypochlorites, etc.

As to the action of these agents on the encrusting materials of retted flax, boiling solutions of caustic and carbonated alkalis saponify and remove the fatty matter, and also decompose the pectic acid and any pectose which may have escaped the action of the retting process. Under their influence the insoluble pectic acid is changed into metapectic acid, which at once unites with the alkali to form a soluble compound. By successive boiling with alkali the fibre entirely loses its brownish colour, and retains only a pale grey shade, readily bleached by hypochlorites. The system of bleaching linen is based on these reactions.

Flax (retted in running or stagnant water) is capable of being well bleached. Under the influence of boiling alkalis it always assumes a lighter colour, and when submitted to the reducing action of stannous chloride it acquires a yellowish tint. Dew-retted flax, on the contrary, bleaches with much difficulty. When boiled with alkalis it becomes darker, and stannous chloride has little or no effect on it. These reactions may serve to discover by which process the fibre has been retted.

Linen fibre is dyed even less readily than cotton, a fact which, although well known to dyers, has not yet been satisfactorily explained. Its physical structure and the possible presence of pectic matters no doubt exercise some restraining influence.

Jute.—This consists of the bast fibres of various species of *Corchorus* (*C. olitorius*, *C. capsularis*, etc.), belonging to the family of the *Tiliaceæ*, and is mainly cultivated in Bengal. The fibre is separated from the plant by processes similar to those employed in obtaining the flax fibre, namely, retting, beating, washing, drying, etc. The raw fibre, as exported, consists of the upper five-sixths of the isolated bast, and occurs in lengths of about 7 ft. Under the microscope it is seen to consist of bundles of stiff, lustrous, cylindrical fibrils, having irregularly-thickened walls, and a comparatively large central opening. The colour of the fibre varies from brown to silver-grey. It is distinguished from flax by being coloured yellow, under the influence of sulphuric acid and iodine solution.

According to Cross and Bevan, the substance of the jute fibre is not cellulose, but a peculiar derivative of it, to which the name bastose has been given. Under the influence of chlorine, a chlorinated compound is produced, which, when submitted to the action of sodium sulphite, develops a brilliant magenta colour. This colour reaction is also exhibited by tannin-mordanted cotton, with which jute shows great similarity; this is further exemplified by the fact that jute can be readily dyed in a direct manner with basic colouring matters.

Jute may be considered as consisting of cellulose, a portion of which has become more or less modified throughout its mass into a tannin-like substance. Alkalis actually resolve jute into insoluble cellulose and soluble bodies allied to the tannin matters. Further, when large masses of jute are allowed to lie in a damp state, the substance of the fibre is decomposed into two groups of bodies, namely, acids of the pectic class, and tannin-like substances.

Acids, notably mineral acids, even at low temperatures, readily disintegrate jute, resolving it into soluble substances. This destructive action of acids must be specially borne in mind by the dyer and bleacher of jute. Strong solutions of hypochlorites produce the chlorinated compound above alluded to, and there is then always a danger of the fibre being disintegrated by subsequent manufacturing operations, as steaming. Weak solutions bleach the fibre to a pale cream colour, at the same time oxidising it, and thus forming compounds which precipitate soluble calcium salts. In bleaching jute, weak sodium hypochlorite should be used in preference to ordinary bleaching powder (calcium hypochlorite), since the presence of soda prevents the formation both of the chlorinated fibre and of insoluble calcium compounds. By thoroughly impregnating the bleached fibre with sodium bisulphite, and drying at 80°-100° C., the colour is still further improved through the action of the disengaged sulphurous acid; the neutral sodium sulphite remaining in the fibre prevents its oxidation and disintegration under the influence of ordinary atmospheric conditions, and even steaming. Jute is readily bleached by the successive action of permanganates and sulphurous acid. The loss of weight experienced by jute in bleaching may vary from 2-8 %, according to the method employed. The standard of moisture is 13·75 %.

China Grass.—This fibre, also called Rhea, Ramie, etc., consists of the bast cells of *Boehmeria nivea* (*Urtica nivea*), a perennial shrub belonging to the nettle family, *Urticaceæ*. The plant grows abundantly in India, China, Japan, and the Eastern Archipelago generally. The native methods of splitting and scraping the plant stems, steeping in water, etc., are tedious and expensive, while the ordinary retting process is not thoroughly effective, because of the succulent nature of the stem, and the large amount and acridity of the gummy matters, which rapidly coagulate and become insoluble on exposure to air. There are now two or three mechanical methods of decorticating the fibre which promise success. The English processes consist of mechanical scrapers, while French experimenters are more in favour of high pressure kier treatment.

The chief characteristics of the fibre are its excessive strength and durability, fineness, silky lustre, and pure white colour. Sulphuric acid and iodine solution colour it blue, hence it seems to consist essentially of cellulose. Under the microscope the fibres appear stiff and straight, the cell walls exhibiting a fibrous texture, and varying in thickness in different parts of the fibre.

CHAPTER III.

WOOL.

Varieties of Wool.—The term wool describes the hairy covering of several species of mammalia, more especially that of the sheep. It differs from hair, of which it may be regarded as a variety, by being, as a rule, more flexible, elastic, and curly, and because it possesses certain details of surface-structure which enable it to be more readily matted together.

Many mammalia have both wool and hair, and it is probable that this has also been the case with the sheep in its original wild state, but under the influence of domestication the rank hairy fibres have largely disappeared, while the soft under-wool round their roots has been singularly developed. The sheep was domesticated at so early a date—as remote, indeed, as the prehistoric period of the Cave Dwellers—that it has been found impossible to determine with certainty its true origin. By some the parent stock is considered to be the *Ovis ammon* of the mountains of Central Asia, where the tribes have always been pastoral in their habits and occupations. The climate, breed, food, and rearing of the sheep, all influence the quality of the wool. When they are fed upon herbage grown in chalky districts, for example, the wool is apt to be coarse, whereas it becomes fine and silky on those reared upon a rich loamy soil.

Sheep's wool varies from the long, straight, coarse hair of certain varieties of the English sheep (Leicester, Lincolnshire, etc.), to the comparatively short, wavy, fine soft wool of the Merino or botany.

Down to the end of the 18th century the Spanish Merino sheep yielded the finest and best wool in the world. About that period this variety was imported into nearly every country in Europe, and by careful selection and good breeding, the wool-growers of Saxony and Silesia at length succeeded in producing wool which quite equalled that of the original Spanish race. Merino sheep have since been introduced into Australia, the Cape of Good Hope, New Zealand, etc.; and these so-called Colonial wools, so much used and appreciated at the present time, all bear the Merino character derived from the original Spanish stock.

According to the average length of the fibres comprising the locks of wool, or "staple," two principal classes of wool may be distinguished, namely, the long-stapled (18-23 cm. = 7-9 in.), and the short-stapled wools (2·5-4 cm. = 1-1·6 in.) In the process of manufacture into cloth, the former require to be combed, and serve for the

Fig. 7.—Microscopical Appearance of Wool Fibre.

production of so-called "worsted" goods, while the latter are carded, and used for "woollen" goods. This distinction between worsted and woollen goods refers rather to the operations of combing and carding than to the length of staple of the wool employed, since large quantities of worsted are made from short wool. The essential difference is really owing to a different arrangement of the fibres in the yarn. The diameter of the wool fibre may vary from 0·007-0·5 mm. = ·002-·02 in.

Very marked differences exist even in the wool of the same animal, according to the part of the body from which it is taken, and it is the duty of the wool-sorter to distinguish and separate the several qualities in each fleece.

The Physical Structure of wool fibre is very characteristic, and enables it to be readily distinguished from other textile fibres. Being a product of the epidermal layer of the skin, it is built up of an immense number of epithelial cells. When carefully examined under the microscope, a wool fibre is seen to consist of at least two parts, sometimes even of three.

(1) The external cells appear as thin horny plates or scales of irregular shape; they are arranged side by side and overlapping each other, somewhat after the manner of roof-tiles (Fig. 7). The upper edges are more or less free, the lower are apparently imbedded in the interior of the fibre. In merino wool the scales appear funnel-shaped, and fit into each other, each one entirely surrounding the fibre. In hair they are more deeply imbedded; they also lie flatter, and present but little free margin.

This surface-character plays an important part in causing the "felting" of wool in "milling," etc. During this and similar operations, in which a large number of fibres are brought into close and promiscuous contact, each fibre naturally moves more readily in one direction than in the other, and the opposing scales gradually become interlocked.

(2) The cortical substance of the wool fibre constituting nearly, and sometimes entirely, the whole internal portion of the fibre, is composed of narrow spindle-shaped cells, which have assumed a more or less horny character. This structure, which gives the inner portion of a wool fibre a fibrous appearance when examined longitudinally under the microscope, is best seen after gently heating the fibre with sulphuric acid. By means of a pair of dissecting needles it is then readily separated into its constituent cells.

In Fig. 8 B represents the microscopic appearance of the fibre after treatment with acid, and A shows some of the individual cells.

It is an interesting fact that these disintegrated internal cells possess a greater attraction for colouring matter than the external scales, and the beneficial effect of the acidity of the bath, required in many cases of mordanting and dyeing, may possibly be ascribed to the opening out of the epithelial scales and the exposure of the inner fibrous cells to the action of the liquid of the bath.

(3) The central, or medullary, portion of the wool fibre, when present, is formed of several layers of rhombic

or cubical cells, which appear as the marrow or pith of the fibre, and may traverse its whole length or appear only in parts. By boiling the fibre with alkali, it is often possible to squeeze out this medullary portion. In many classes of wool (merino, etc.) it seems to be entirely absent: indeed, its presence or absence depends upon a variety of factors, as race, health of sheep, part of the body from which the wool is taken, etc.

In colourless wool the medullary portion often appears under the microscope as a dark or dull hazy stripe. This appearance is caused by the air enclosed between the cells,

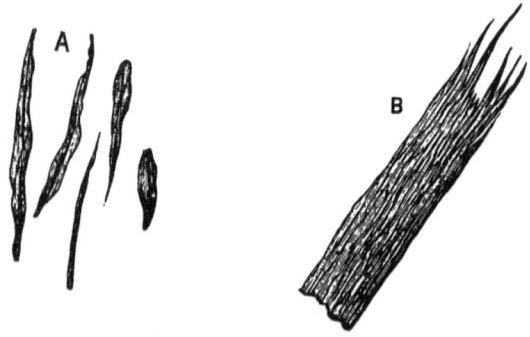

Fig. 8.—Cells of Wool Fibre under the Microscope.

so that by boiling such a fibre with turpentine or glycerine the cells become transparent.

Wool fibres which exhibit the medullary cells are brittle and stiff; altogether they possess more of a hairy character, and are less suitable than other varieties for manufacturing purposes. In the best qualities of wool the medullary cells are invisible.

In a transverse section a wool fibre appears more or less round or oval.

Fig. 9 gives the cross sections, according to F. Bowman, of two typical wool fibres; A shows the medullary, cortical, and external cells; in B the medullary cells are absent.

"Kemps," or dead hairs, are certain wool fibres not possessing the normal structure of good wool; under the microscope the epithelial scales are less distinct, or even invisible, and viewed by transmitted light, either the whole substance of the fibre seems more dense and sometimes even opaque, or the medullary portion only is opaque. They are deficient in tenacity, lustre, and felting power, and in their attraction for colouring matters. They may occur even in good qualities of wool, about the neck and legs of the animals. In coarse wools they may be found in any part of the fleece.

A merino wool fleece is made up of an immense number of small bundles or strands of wool fibres, which, in the best races of sheep, show a perfectly regular and fine wavy character. The individual fibres are also more or less wavy, but not with the same degree of regularity as the strand of which they form a part. When the fibres

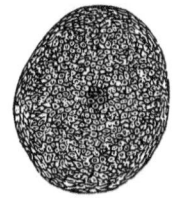 

Fig. 9.—Cross Section of Typical Wool Fibres.

adhere to each other, as in the strand, the regular wavy character is very marked.

The hairy covering of animals other than sheep's wool is used in the woollen industry.

Foreign Wools.—Alpaca, Vicuna, and Llama wool are obtained from different species of the genus Auchenia (*A. alpaca, A. vicugnia, A. llama*), which inhabit the mountains of Peru and Chile.

Mohair is obtained from the Angora goat (*Capra hircus angorensis*) of Asia Minor.

Cashmere consist of the soft under-wool of the Cashmere goat (*Capra hircus laniger*) of Tibet.

The soft under-wool of the camel, which it sheds each spring, is also used. Of all these, the alpaca and mohair are most largely employed.

Certain of these foreign wools, more especially Van Mohair, also Alpaca, Camel's hair, Cashmere, and Persian wool, are apt to be dangerous to the health of the woolsorter. They seem to contain the microscopic organism known as *Bacillus anthracis*, the same which excites splenic fever in cattle and horses. When taken into the bronchial tubes of man, it induces a kind of blood-poisoning known as " wool sorter's disease." Although once very common, this disease is now rare, for there are stringent regulations as to the boards and rooms where dangerous wools are sorted.

Hygroscopicity of Wool.—Wool fibre is capable of absorbing a large amount of water without appearing damp, that is, it is very hygroscopic. Exposed to the air in warm, dry weather, it contains 8-12 % moisture; but if kept for some time in a damp atmosphere, it may take up as much as 30-50 %. This moisture probably fills up the interstices between the cells of the fibre, which under ordinary circumstances contain air, but it no doubt also permeates the substance of the cells themselves. It is noteworthy that damp wool is not so liable to mildew as the vegetable fibres are.

The amount of moisture in unwashed wool varies with the fatty matter it contains, the less fat the more moisture; while in washed wool it depends upon the arrangement of the cells. The wool which has least tenacity—that is, that in which the cells are more loosely arranged—possesses the greatest hygroscopicity.

This hygroscopic character of wool renders it very desirable that those trading with it should know exactly its condition in this respect at the time of buying and selling, hence conditioning houses have been established in this country and on the Continent, where the exact amount of moisture in any lot of wool may be officially determined. The legal amount of moisture allowed on tops in the oil is 19 %, on tops combed without oil and worsted yarn $18\frac{1}{4}$ %, and on woollen yarn 17 %.

If wool fibre is steeped in warm water, it softens and swells very considerably, and, like all horny substances, becomes plastic, retaining any position which may be forced upon it, if, while the mechanical strain is continued, the moisture is more or less evaporated and the temperature reduced.

This hygroscopic and plastic nature of wool comes

into play in the processes of "crabbing" and "steaming" of unions, in the "boiling" and "finishing" ("hot-pressing") of woollen-cloth, and in the "stretching" of yarn.

Elasticity of Wool.—Closely connected with the hygroscopic nature of wool is its elasticity, which it possesses in a high degree, not merely because of the wavy character of the fibre, but also on account of its substance and structure. One important manifestation of its elasticity is shown if a dry wool fibre is excessively stretched; when the ends are released, or rupture takes place, the fibre or the separated parts rebound to the original position, and an additional shrinking and curling up of the ends are exhibited.

If a single wool fibre is softened by heat and moisture, then stretched and dried in this condition, it is found to have lost this curling property, but it reappears whenever the stretched fibre is again softened, and allowed to dry in an unfettered condition.

In conjunction with pressure, friction, and temperature, many of the above-mentioned physical features—such as the scaly surface of the fibre, its waviness, and its hygroscopic, elastic, and plastic nature—play a most important part in the processes of "felting" and "milling" woollen cloth.

The lustre of wool varies very considerably. Straight, smooth, stiff wool has more lustre than the curly merino wool. The differences exhibited depend partly upon the internal structure, but chiefly upon the varying arrangement and transparency of the scales on the surface of the fibre; the flatter these are and the more they lie in one plane, the greater will be the lustre. Wools such as Lincoln and Leicester, etc., which possess a silky lustre in a high degree, are classed as lustre wools, as distinguished from non-lustre wools, such as Merino, Colonial, etc.

Wool with a glassy lustre, such as bristles, etc., is harder and more horny than non-lustre wool; the surface is smoother, the scales are less distinct, and wools of this kind do not dye so readily.

The best kind of wool is colourless, but lower qualities are often yellowish, and sometimes variously coloured, such as black, brown, red, etc. This coloration is caused by the presence of an organic pigment in the cortical portion of the fibre, either as a granular pigment situated between the cells, or as a colouring matter diffused throughout the cell substance. Generally, both forms are present, but in brown and black wool the granular pigment predominates, while in red and yellow wools the diffused colouring matter is more prominent. These natural pigments are not so fast to light as is generally supposed, a fact which is revealed by the bleached appearance of the exposed portions of the fleece.

The worth of any quality of wool is determined by carefully observing a number of its physical properties, such as softness, fineness, length of staple, waviness, lustre, strength, elasticity, flexibility, colour, and the facility with which it can be dyed. Fleece wool, as shorn from the living animal, is superior in quality to " dead wool," that is, wool which has been removed from the skin after death, if lime has been used in the process, but if it be removed from the skins by cutting, the wool is practically equivalent to " fleece wool "; indeed, it is said to felt better than the latter. Individual dead fibres occur occasionally in fleece wool; they have been forced out by the roots previous to the time of shearing, and constitute the so-called " overgrown " wool. This wool is comparatively harsh and weak, and does not dye so readily as other kinds. This is the case also with the wool of an animal which has died of some distemper.

Chemical Composition of Wool.—A distinction must be made between the fibre proper and the foreign matters encrusting it. The latter, though consisting partly of mechanically adhering impurities derived from without, are mainly secreted by the animal, and constitute the so-called *Yolk* (Fr. *Suint*). It may be remarked that it is the estimation of these foreign matters which makes wool-buying so difficult and necessitates both experience and quick judgment to prevent later loss.

Wool fibre which has been entirely cleansed and freed from these foreign matters possesses a chemical composition very similar to that of horn and feathers, and consists of what is termed Keratin (horn-substance). Its elementary composition varies somewhat in different qualities of wool, but the following analysis of German wool may be taken as representative:—

Carbon	49·25%
Hydrogen	7·57%
Oxygen	23·66
Nitrogen	15·86
Sulphur	3·66
	100·00

Whether the sulphur is an essential constituent or not is a question that has been much discussed. It is removed to a greater or less degree by most solvents, hence it is difficult to obtain constant analytical results. Its amount has been found to vary in different wools from 0·8-3·8 %. Its constant occurrence, and that in comparatively large proportion, precludes the idea that it is merely an accidental constitutent, and it has hitherto been found impossible to deprive wool entirely of its sulphur, without, at the same time, modifying somewhat its structure and in large measure destroying its tenacity.

This presence of sulphur in wool is attended with some practical disadvantages. The wool is apt to contract dark-coloured stains under certain conditions, and on that account its contact with such metallic surfaces as those of lead, copper, and tin should be avoided during processes of scouring or dyeing. In mordanting with stannous chloride and cream of tartar, especially if an excess of these ingredients be used, the wool is frequently stained, by reason of the formation of stannous sulphide.

A boiling solution of plumbite of soda at once blackens wool, and may thus serve to distinguish it from silk or cotton.

For practical purposes, much of the sulphur may be removed by steeping the wool in cold weak alkaline solutions, such as milk of lime—then washing it in water, in weak hydrochloric acid, and again with water, repeating the operations several times.

The amount of mineral matter in wool free from yolk varies from 0·08-0·37 %. It consists mainly of phosphates and silicates of lime, potash, iron, and magnesia.

Action of Heat on Wool.—Heated to 130° C., wool begins to decompose and give off ammonia; at 140-150° C. vapour containing sulphur is disengaged.

Wool fibre inserted in flame burns with some difficulty, and emits a disagreeable odour of burnt feathers. It has the appearance of fusing, a bead of porous carbon being formed at the end of the fibre. Submitted to dry distillation, it gives off products containing much ammonium carbonate, which may be readily detected by its smell or by its colouring red litmus-paper blue. These reactions serve to distinguish wool from all vegetable fibres.

A cold ammoniacal solution of cupric hydrate has no action upon wool, but if it is used *hot* the wool is dissolved.

Action of Acids on Wool.—Dilute solutions of hydrochloric and sulphuric acids, whether applied hot or cold, have little influence upon wool, further than opening out the scales and making the fibre feel somewhat rougher, but if used concentrated, the fibre is soon disintegrated; in any case their destructive action is by no means so energetic on wool as on cotton. This fact is made use of to separate cotton from wool in the process of "extracting" or "carbonising" rags containing both fibres. The rags are steeped in dilute sulphuric acid, and after removing the excess of liquid, are dried in a stove at about 110° C. The disorganised cotton can then be beaten out as dust, while the wool remains comparatively little injured. Another method is to submit the rags for a few hours to heated hydrochloric acid gas. The above mineral acids are frequently added to the dye-bath in wool-dyeing.

Nitric acid acts like the acids just mentioned, but it also gives a yellow colour to the wool, owing to the production of so-called xanthoproteïc acid. Because of the comparatively light yellowish colour thus imparted, boiling dilute nitric acid is frequently used as a "stripping" agent for wool—to destroy the colour in wool already dyed—for the purpose of re-dyeing (job-dyeing, rectifying mistakes, etc.). Care must always be taken not to have the acid too strong (about 3°-4° Tw.—Sp. Gr. 1·02), and not to prolong the process beyond three or four minutes.

Sulphur dioxide (sulphurous acid gas) removes the natural yellow tint from ordinary wool, and is the best bleaching agent employed for this fibre. It is important to remember that the gas is very persistently retained by the fibre, and should always be removed from bleached wool previous to dyeing light colours. This is effected by steeping the wool in very dilute solutions of carbonate of soda or bleaching-powder, and washing well. When the first reagent is employed, the acid is merely neutralised, but with the second the sulphurous acid is oxidised to sulphuric acid. Should this precaution be neglected, the wool will not dye properly, or, when dyed, it will be liable to become decolorised again through the reducing action of the sulphur dioxide retained by the fibre.

Action of Alkalis on Wool.—Alkaline solutions have a very sensible influence on wool, but the effects differ considerably according to the nature of the alkali, the concentration and temperature of the solution, and the duration of contact.

Caustic alkalis, (KHO, NaHO) act injuriously on wool under all circumstances. Even when they are applied as cold and weak solutions, their destructive action is sufficient to warrant their complete rejection as " scouring " agents.

When they are applied hot, even though but little concentrated, the wool is gradually dissolved, producing a soapy liquid from which it may be precipitated, on the addition of acid, as a white amorphous mass.

This fact of the solubility of wool in hot caustic alkalis is utilised for the purpose of recovering indigo from vat-dyed woollen rags, this colouring matter being insoluble therein.

Solutions of alkaline carbonates and of soap have little or no injurious action on wool, if they are not too concentrated, and the temperature is not higher than 50° C. Soap and carbonate of ammonia have the least injurious action, while the carbonates of potash and soda impart to the wool a yellow tint, and leave it with a slightly harsher and less elastic feel.

This marked difference of action between the caustic and carbonated alkalis makes it an all-important matter for every wool-scourer to know the exact nature of the agents he uses soaps should be free from excess of alkali, " soda ash " should contain no caustic soda, etc.

Calcium hydrate (lime) acts injuriously, like the caustic alkalis, but in a less degree. It eliminates the sulphur from the wool, but thereby renders the fibre brittle and impairs its milling properties.

Action of Chlorine and Hypochlorites on Wool.—These

act injuriously, and can therefore never be applied as bleaching agents. A hot or boiling solution of chloride of lime entirely destroys the fibre, with evolution of nitrogen gas; if, however, wool be submitted to a very slight action of chlorine or hypochlorous acid, it assumes a yellowish tint, and acquires at the same time an increased affinity for many colouring matters. This effect is possibly due to an oxidation of the fibre, and not merely to a roughening of its surface. Practical use is made of it by the printer of Muslin Delaine (mixed fabrics of cotton and wool) and occasionally by the woollen dyer.

Action of Metallic Salts on Wool.—In common with all fibres of animal origin, wool has the property of readily dissociating certain metallic salts when in contact with their solutions, especially if the latter are heated. When, for example, wool is boiled with solutions of the sulphates, chlorides, or nitrates of aluminium, tin, copper, iron, chromium, etc., a small amount of the oxides or of insoluble basic salts of these metals is deposited upon or attracted by the fibre, and a more acid salt remains in solution. On this fact depends the method of mordanting wool, which differs from that employed with the vegetable fibres, since these do not cause dissociation under like conditions. Neutral salts of the alkalis (such as $NaCl$, Na_2SO_4) exercise no appreciable action on wool.

Action of Colouring Matters on Wool.—Wool has a marked direct attraction for certain colouring matters (magenta, azo-scarlet, indigo extract, orchil, etc.) if their solutions are presented to it in a proper state of neutrality or acidity, etc., and with these it is dyed with great facility. Being of a porous nature, it is indeed readily permeated by solutions of all colouring matters, especially when heated with the latter.

Yolk in Raw Wool.—The foreign matter, or yolk, enveloping the pure wool fibre possesses a special interest for the dyer, because on its entire removal depends to a very large extent the success with which he may obtain fast, pure, and even colours. To the merchant and manufacturer it is also of great importance, since the amount in different kinds of raw wool varies considerably, and influences its commercial value.

By treating wool first with distilled water, and afterwards with alcohol, Chevreul obtained the following analysis of raw merino wool dried at 100° C. :—

Removed by water.	Yolk soluble in cold distilled water	32·74%
	Earthy matter deposited from the above	26·06%
Removed by alcohol	Fatty matter dissolved by alcohol	8·57%
	Earthy matter adhering to the fat	1·40%
	Wool fibre	31·23%
		100·00

Chevreul has here designated as yolk only those impurities which are soluble in cold water, although in the ordinary commercial acceptation of the term it includes all adhering impurities.

According to other observers, the amount of the main constituents of "raw" or "greasy" wool, just as it comes from the sheep's back, may vary considerably, according to its origin, as follows :—

Moisture	4-24%
Yolk	12-47%
Wool fibre	15-72%
Dirt	3-24

As a rule, the finer qualities of wool, such as merino, contain more yolk than the coarser.

When wool is washed with water, as in Chevreul's analysis, not only are certain constituents of a soapy nature —*i.e.* alkaline oleates—removed, but some of the unsaponifiable matter as well, since the oleates cause it to form an emulsion.

A better method of separating these two constituents is to treat the dried raw wool first with ether. This dissolves principally the fatty matter, and although it also takes up as much as 10 % of the oleates present, repeated washing of the ethereal solution with water removes the latter almost entirely.

The following substances may thus be distinguished in raw wool—wool-fat (soluble in ether), wool-perspiration (soluble in water, and partly also in alcohol), wool fibre, dirt, moisture.

These may be determined as follows :—

(*a*) Weigh the raw wool, dry it at 100° C., preferably in a stream of some dried inert gas—as hydrogen—and weigh again. The loss in weight gives the moisture present.

(b) Extract the dried wool with ether, shake up the ethereal solution with water, in order to remove from it the oleates; evaporate the separated ether to dryness, and weigh the fatty residue. The weight gives the amount of wool-fat present. Evaporate the separated wash-water to dryness, weigh the residue, and add the weight to that of the portion soluble in water : the oleates.

(c) Wash the ether-extracted wool several times with cold distilled water, and evaporate the solution to dryness. The weight of the residue added to the weight of the oleates dissolved by water from the ethereal solution gives the chief amount of the alkaline oleates present. The wool is then washed with alcohol; this always dissolves further minute quantities of oleates, the weight of which must be added to the above. Earthy oleates which remain in the wool are decomposed by washing the latter with dilute hydrochloric acid; the acid is removed by washing with water, the wool is then dried, and extracted with ether and alcohol. From the weight of the residue obtained on evaporating the two last solvents to dryness the amount of earthy oleates present in the wool may be calculated. With very dirty wool a good deal of lime is dissolved by the hydrochloric acid, not because of lime soaps but of calcareous dust present.

(d) The wool remaining is dried and thoroughly well shaken and teazed out by hand over a large sheet of paper, in order to remove dirt, sand, etc. ; care is taken not to lose any of the fibre, the detached particles of which are collected on a fine sieve, and washed with water till free from dirt. The wool is dried and weighed; the sand, dirt, etc., are determined by difference.

The following analyses of raw wools give the results obtained by the above method of Märcker and Schulz :—

	Lowland Sheep.	Rambouillet Sheep	Pitchy Wool.
Moisture (per cent.)	23·48	12·28	13·28
Wool-fat	7·17	14·66	34·19
Soluble in water (wool-perspiration)	21·13	21·83	9·76
Soluble in alcohol	0·35	0·55	0·89
Soluble in dilute HCl	1·45	5 64	1·39
Soluble in ether and alcohol	0·29	0·57	—
Pure wool fibre	43·20	20·83	32·11
Dirt	2·93	23·64	8·38

(By successive treatment.)

WOOL. 43

Wool-fat.—The composition of what is here called wool-fat is found to be of a somewhat complicated nature. By treating it with boiling alcohol it may be separated into two portions, the one soluble, the other and larger amount insoluble in this liquid. Further analysis has shown that the soluble portion consists mainly of the alcoholic and fat-like body cholesterine, together with isocholesterine, each in the free state, and probably also of compounds of both these bodies with such organic acids as acetic acid. The insoluble portion consists essentially of compounds of cholesterine and isocholesterine with oleïc acid, and in less amount with solid fatty acids, such as stearic acid and hyæna acid.

There seems also to be present in a similar state of combination, but in smaller quantity, some other amorphous body or mixture of bodies, readily fusible, of an alcoholic nature, and containing less carbon than cholesterine. A portion of these various alcoholic bodies, and sometimes also a part of the high atomic fatty acids, are present in the free state.

Wool-fat is certainly not a compound of glycerine, and hence is not a fat as ordinarily understood. This accounts for the difficulty experienced in removing it by mild scouring agents from so-called " pitchy wool," which, as shown in the above analysis, contains it in excessive quantity.

Wool-perspiration.—With reference to the chemical composition of that portion of the yolk which is soluble in water, the wool-perspiration, it has been shown by the experiments of Vauquelin, Chevreul, Hartmann, and others, that it consists essentially of the potassium compounds of oleïc and stearic acids, and probably also of other fixed fatty acids; it contains further, but in smaller amount, the potassium salts of certain volatile fatty acids (acetic and valerianic acid), potassium chloride, phosphates, sulphates, etc.

Ammonium salts seem to be present in dried extracted yolk in small quantity (equivalent to 0˙5 % NH_3), not sufficient, however, to account for the amount of nitrogen found in yolk (3 %); some other nitrogenous body is evidently present.

As a rule, the wash-water of raw wool has a strong

alkaline reaction, since potassium carbonate may be present to the amount of 4 % of the weight of raw wool. Some observers have found it to be entirely absent, in which case, however, it may still be considered to have been secreted by the perspiration glands of the sheep, but to have afterwards acted energetically upon the wool-fat and saponified it, so that while disappearing itself, it has given rise to an increased amount of potash-soaps in the wool-perspiration.

In washing wool on the sheep's back, this potassium carbonate, when present, plays a not unimportant part along with the potash soaps of the yolk, in greatly facilitating the removal of dirt, etc., from the fleece.

Dried extracted yolk contains about 60 % organic matter and 40 % mineral matter (free from CO_2).

The following are two analyses of yolk-ash by Märcker and Schulz:—

Potash	58·94%	63·45%
Soda	2·76%	trace
Lime ...	2·44%	2·19%
Magnesia	1·07%	0·85%
Ferric oxide ...	trace	trace
Chlorine ...	4·25%	3·83%
Sulphuric acid	3·13%	3·20%
Phosphoric acid	0·73%	0·70%
Silicic acid ...	1·39%	1·07%
Carbonic acid	25·79%	25·34%

Yolk-ash consists essentially, therefore, of potash salts, principally carbonates, the carbonic acid arising mainly from the burning of the organic constituents of the yolk.

Maumené and Rogelet give the following analysis, which closely agrees with the above:—

Potassium carbonate	86·78%
Potassium chloride	6·18%
Potassium sulphate ...	2·83%
SiO_2, P_2O_5, CaO, MgO, Al_2O_3, Fe_2O_3, Mn_2O_3, CuO	4·21%
	100·00%

It is evident from the above that when wool is washed on the sheep's back a considerable quantity of potash is entirely lost to the farmer. It has been estimated that raw wool yields 8·75 % by weight (free from CO_2), and if the nitrogenous matter and phosphates also washed away are taken into account, it will be seen that the wash-water of raw wool posesses an appreciable manurial value.

The Wash-Water Products of Raw Wool.—The great bulk of the commercial wool is in the unwashed or "greasy" condition, so that an opportunity is afforded to the woollen manufacturer of extracting the whole of the yolk, and making it serve as a supplementary source of potash.

It is interesting to know that since 1860, and based mainly upon the observations of Maumené and Rogelet, the manufacture of potash salts from the wash-water of raw wool, used in the centres of the French and Belgian woollen industry, has become an accomplished fact, the annual production of potassium carbonate being estimated at about 1,000,000 kg.

After systematically washing the wool with water, the saturated solution is evaporated to dryness. The residue is heated in gas retorts, and the gas evolved may be used for illuminating purposes. The resulting coke is either calcined with access of air or lixiviated with water, and yields crude potassium carbonate. "Greasy" wool yields 7-9 % crude potassium carbonate, containing 85 % K_2CO_3.

Another mode of utilising yolk is that recommended by Havrez, according to whom it is the natural raw material for the manufacture of yellow prussiate of potash. The ordinary method of making this salt is to heat a mixture of crude carbonate of potash, waste animal matter (dried blood, leather clippings, etc.), and iron filings. The resulting fused matter is extracted with water, and on evaporating the solution the desired salt is obtained.

Havrez says that when yolk is submitted to dry distillation it yields a residue, which is an extremely intimate mixture of carbonate of potash and nitrogenous carbon. This residual coke contains, therefore, just the necessary elements for the production of yellow prussiate of potash, and experiment has shown that it gives even a greater yield than the ordinary mixture, containing an equal amount of K_2CO_3, because of the perfect and intimate mixture of the various ingredients.

Havrez has calculated that the money value of the

yolk, when used for the production of yellow prussiate of potash, is more than twice that of its ordinary commercial value. He further maintains that when it is used for the simultaneous production of carbonate of potash and yellow prussiate of potash, instead of the former only, there is a gain in value of 50 %. For this purpose the dried yolk is mixed with an equal weight of waste animal matter, and heated somewhat longer than usual. Experiment showed that carbonate of potash obtained from 100 kg. of the residual melt was accompanied by 17·3 kg. of potassium cyanide, which was capable of yielding 19 kg. of yellow prussiate of potash. One hundred kg. of yolk treated in this manner are said to yield 32 kg. of carbonate of potash and 4·3 kg. of yellow prussiate of potash.

Previous to its employment in manufacturing or in dyeing, the raw wool must be thoroughly cleansed from the yolk, but since only a portion is removed by a simple treatment with water, recourse is had to the detergent action of solutions of soap, alkaline carbonates, etc.

No attempt seems yet to have been made in England to collect separately the soluble portion of the yolk for the purpose of recovering the potash salts. The preliminary extraction with water, or steeping, alluded to, is dispensed with by the manufacturer, and the wool is at once washed with solutions mentioned above. This operation is termed "scouring," and will be treated of in detail in a future chapter; but it may be well to state here that, although in the English method the potash salts are entirely lost, the alkaline and detergent properties of the soluble portion of the yolk are utilised.

CHAPTER IV.

SILK.

Origin and Culture of Silk.—Silk differs entirely both from the vegetable fibres and from wool by being devoid of cellular structure. It consists of the pale yellow, buff-coloured, or white fibre, which the silkworm spins round about itself when entering the pupa or chrysalis state. The numerous varieties of silk may be conveniently divided into two classes, cultivated and wild silk. The latter is the product of the larvæ of several species of wild moths, which are natives of India, China, and Japan. The former and more important class is produced by the common silkworm, or caterpillar of the moth *Bombyx mori*, which has become the subject of special culture (see Fig. 12). The chief seats of the silkworm culture are Southern Europe (including the South of France, Italy, and Turkey), Japan, China, and India.

The eggs of the European silk moth are about the size and shape of poppy seeds. One g. weight of them contains about 1,350 eggs. They have at first a yellowish colour, which, however, on drying, changes to grey. The rearing of the silkworm is mainly conducted in specially-arranged establishments, called *Magnameries*. In these, the incubation-chamber is a well-lighted, airy room, where the eggs are spread out on sheets of paper resting on lattice-work. A certain suitable degree of moisture is maintained, and the temperature is gradually raised in the course of about 10-12 days from $18°$-$25°$ C. The young caterpillars, as soon as they appear, are taken to a more roomy chamber, in which there is erected a lath framework strung across with threads and sheets of paper. Here the animals are regularly fed during 30-33 days, till, indeed, they begin to spin. Their food consists of the leaves of the mulberry tree, *Morus alba*, hence the silk is frequently termed mulberry silk. During the feeding period the silkworm increases enormously in size, to about 8-10 cm. (3-4 in. long) and about 5 g. (77 grains) in weight. As might be expected, such a rapid and enormous development necessitates a frequent renewal of the skin, and moulting takes place three or four times, at tolerably regular intervals of 4-6 days. On or about the thirteenth day the animal ceases to take food, and evinces a restless activity. At this period it is placed on birch twigs, etc., where it soon begins to spin. The silk substance is secreted by two

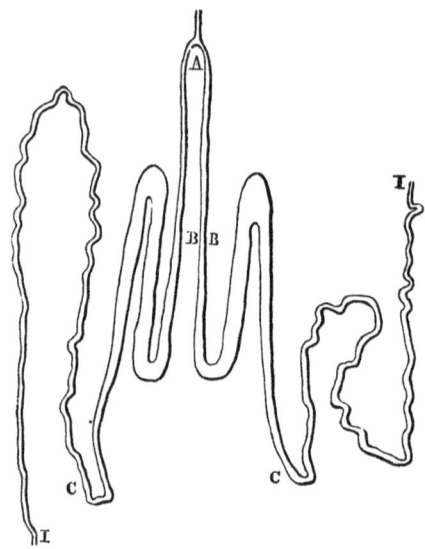

Fig. 10.—The Silk Glands of the Silkworm.

glands symmetrically situated on each side of the body of the caterpillar, below the intestinal canal. Each gland, as shown in Fig. 10, consists of three parts : a narrow tube (I C) with numerous convolutions, the veritable secreting portion; a central part (C B) somewhat expanded, and constituting the reservoir of the silk substance; a capillary tube (B A), connecting the reservoir with a similar capillary canal at A, common to both glands, and situated in the head of the animal, whence issues the silk.

The silk substance as contained in the central reservoir

is a clear, colourless, gelatinous liquid. According to Duseigneur, this is surrounded by a layer of another substance, colourless when the silk is white, coloured when it is yellow, and which possibly constitutes the silk-gum to be alluded to subsequently. The whole is enclosed in a thin membrane. A transverse section (Fig. 11) shows that it occupies a space equal to 20-25 % of the total volume, a

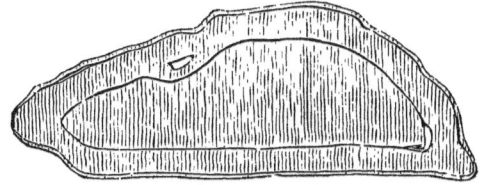

Fig. 11.—Section of Silk-bag.

proportion which corresponds somewhat to the loss sustained by raw silk during the operation of "boiling-off."

Arrived in the capillary tube at A (Fig. 10), the silk substance solidifies, and issues from the spinneret in the form of a double fibre, as represented in Fig. 12. Occasionally the two fibres may be slightly separated at

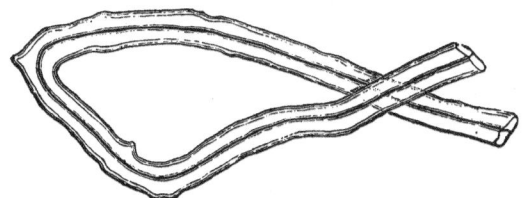

Fig. 12.—Microscopic Appearance of Raw Silk Fibre.

intervals, and form then at these points two transparent solid cylinders.

In the beginning of its spinning operations the silkworm throws round about itself a light scaffolding, as it were, of short fibres connecting the neighbouring points of support. When this is completed its movements become slower, and by moving its head from side to side it gradually forms and lines its dwelling with numerous layers of what may be termed silken lattice-work.

Towards the interior of the layers they become firmer and denser, while the innermost one, which immediately protects the animal, forms a thin parchment-like skin. The egg-shaped product is called a cocoon (Fig. 13). It is made up of a double fibre, only rarely broken, varying in length from 350-1,250 m. (400-to 1,400 yds.), and with a diameter of about 0·018 mm. (0·0007 in.). Each fibre is thickest in the outermost portion of the cocoon, and becomes thinner towards the interior, owing to the exhaustion of the caterpillar from want of food during the spinning process.

The cocoons are white or yellow, contracted in the centre, about 3 cm. = 1·18 in. long, and 1·5-2 cm. = 0·59-0·78 in. thick.

As soon as the metamorphosis of the caterpillar into

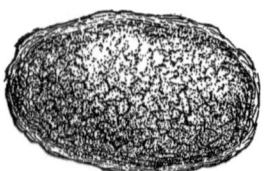

Fig. 13.—Silk Cocoon.

the chrysalis state is completed, the cocoons are collected. Those which are intended for breeding purposes are left to themselves in a room heated to 19°-20° C. Three weeks after the spinning of the cocoon, the silk moth, which has now been formed in the interior, emits a peculiar kind of saliva; with this the animal softens one end of the cocoon, and pushes its way out. A few days after the females have laid their eggs they die, not being provided with any organ of nutrition. The eggs are slowly dried, and stored in glass bottles in a dry dark place till the following spring.

Experience has shown that the worms issuing from 100 g. of eggs consume 3,500-5,000 kg. = 3½-5 tons of leaves, and produce, under favourable conditions, 87,900-117,200 cocoons, weighing 150-200 kg. = 330·7-440·9 lb., and these yield 12-16 kg. = 26·5-35·3 lb. of reeled silk.

In their natural state cocoons contain generally :—

Moisture	68·2%
Silk	14·3%
Floss (*bourre*)	0·7%
Chrysalis	16·8%

The good silk is obtained from those cocoons of which the pupæ are killed, either by heating the cocoons for 2-3 hours in an oven heated to 60°-70° C., or by means of steam. The latter method is the more general, the apparatus consisting essentially of a boiler for generating the steam, and a steam-box in which the cocoons are placed. When subjected to steam, the pupæ are killed in 10-12 minutes.

The external loose flossy silk is first removed, and the cocoons are thrown into baskets, which are then placed in the steaming-box. When taken out, the cocoons are put between woollen blankets, so that the heat may be retained and its effect continued a little longer. After a few hours they are spread out on tables, and shovelled about till perfectly dry. This last operation is specially requisite with such cocoons as cannot be reeled off at once. Although the killing of the cocoons by steam has the great advantage that there is no fear of the silk itself being damaged by overheating, still it has certain defects. The most serious are, that some of the pupæ burst and soil the silk, and that the fibres soften somewhat, and tend to stick together, rendering the subsequent reeling more difficult.

After killing, the cocoons are "sorted," or divided into classes of different quality. In every piece of woven silk the warp threads have to bear the greatest strain, and, as a rule, must appear on the surface of the fabric, hence the best cocoons are chosen for the warp, since they must yield strong, smooth, even, and lustrous fibres. These fibres, too, in the subsequent process of reeling, are manipulated somewhat differently from the weft-fibres. The product of this choice and particular treatment forms the best quality of silk, that is, warp-silk, which is known as Organzine. A somewhat inferior class of cocoons is worked up to form weft-silk, or Tram.

Raw Silk.—In the reeling process a number of cocoons (4-18) are thrown into a basin of warm water, in order to soften the gummy envelope of the fibres, thus permitting their ready separation from the cocoon, and also to cause the subsequent agglutination of whatever number of fibres may be thrown together to form a single thread. During this reeling process two threads, composed of an equal number of fibres, are passed separately through two perforated agate guides; after being crossed or twisted together at a given point, they are again separated, and passed through a second pair of guides, thence through the distributing guides on to the reel. The object of this temporary twisting or crossing (Fr. *croissage*) is to cause the agglutination of the individual fibres of each thread, and to aid in making the latter smooth and round. The unequal diameter of each fibre at different portions of its length is taken into account by the reeler when introducing new fibres into the thread to replace those which have run out. The quality of raw-silk depends very much indeed upon the care bestowed on the reeling process.

The loss through removal of the external floss (*bourre*) varies from 18-30 %, according to the cocoons and the care bestowed by the worker.

Reeled-silk or raw-silk, as it is generally termed, constitutes the raw material of the English silk manufacturer. Before being used for weaving, two or more of the raw-silk threads are " thrown " together and slightly twisted by the silk spinner, or " throwster," and in this way the various qualities of Organzine and Tram—also embroidery-sewing silks, etc.—are produced.

A very brief description of the operations involved may suffice. The preliminary processes of " winding " and " cleaning " are followed by that of " doubling," which simply consists in placing two threads (" singles ") side by side, and winding them together without twist. The so-called " spinning " of silk consists merely in twisting the threads either before or after doubling.

The " Tram," already alluded to, is the product of the union of two or more single untwisted threads, which are then doubled and slightly twisted.

" Organzine " is produced by the union of two or more single threads separately twisted in the same direction, which are doubled and then re-twisted in the opposite direction.

Waste-silk.—This proceeds from perforated and double

cocoons, and such as are soiled in steaming; also the extreme outer and inner portions of the cocoons in short, all silk obtained from cocoons in any way soiled or unable to yield a continuous thread. The killed chrysalides can be used as a source of oil, and the residue after extraction may serve for manure.

All such waste-silk materials are washed, boiled with soap, and dried. They are afterwards carded and spun somewhat after the manner of cotton and flax, and yield the so-called Spun silk. Schappe silk is a similar product, made from waste silk without previous boiling.

Careful examination of the perforated cocoons has revealed the fact that their fibres are not discontinuous. The moth does not eat its way out of the cocoon, but rather pushes aside the previously softened fibres. Attempts made to reel these cocoons by the ordinary methods have failed, since the cocoons soon fill with water and sink.

Wild Silk.—The most important is the Tussur silk (Hindustani, *Tusuru*, a shuttle), also called Tusser, Tasar, Tussore, and Tussah. It is the product of the larva of the moth *Antherœa mylitta*.

The Tussur moth is found in nearly all parts of India, where it seems to feed on a large variety of plants. In some districts it is reared by the natives. The cocoons, which are much larger than those of *Bombyx mori*, are egg-shaped and of a silvery drab colour. They are attached to the twigs of the food-trees by a peduncle having a terminal ring. The outer silk is somewhat reddish, and consists of separate fibres of various length, while the rest is generally unbroken to the centre of the cocoon. The cocoon is extremely firm and hard, the fibres being cemented together by a peculiar secretion of the animal, which permeates the whole wall of the cocoon, and imparts to it its drab colour. The cocoons are boiled and carded, or even reeled, although this latter process presents difficulties. Silk plush is largely made from carded Tussur silk.

Other wild silks are, Eria silk, from *Attacus ricini;* Muga silk, from *Antherœa assama;* Atlas silk, from *Attacas atlas;* Yama-maï silk, from the *Antherœa yama-maï* of Japan, etc.

Under the microscope a raw mulberry-silk fibre appears as a double fibre (Fr. *bave*) consisting of two solid structureless cylinders (Fr. *brin*), more or less united together (Fig. 12); after "boiling-off" with soap, however, this double fibre separates into a pair of distinct fibres, having a more or less irregular, somewhat rounded triangular section.

Wild silk is distinguished from mulberry silk by the longitudinal striations seen in each of the double fibres when under microscopic examination (Fig. 14), and by the apparent contraction of the fibre at certain points. The former are due to the fact that the wild-silk fibre is composed of a large number of fibrils, while the latter appear-

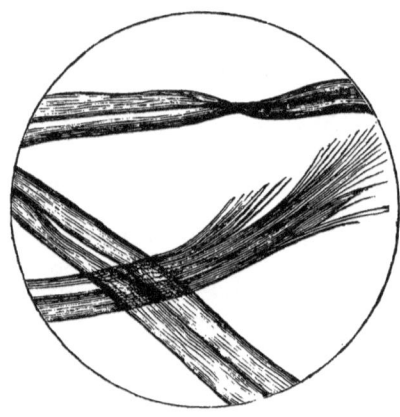

Fig. 14.—Microscopic Appearance of Tussur Silk Fibre.

ance is seen because the more or less flattened fibres are twisted at the contracted points.

Physical Properties of Silk.—The most important are its lustre, strength, and avidity for moisture. One other distinctive property which it possesses in certain conditions is that of emitting a peculiar crisp, crunching sound (Fr. *cri*) when a bundle of silk yarn is tightly twisted and pressed together. This peculiar property is called the "scroop" (Fr. *craquant*) of the silk, and no doubt gives rise to the rustling noise heard when two pieces of silk fabric rub lightly against each other. The intensity of the phenomenon varies with the nature of the dyeing and mechanical processes adopted, with the diameter and twist of the threads, etc.

The property is absent in raw-silk and in "boiled-off," or in "souple" silk; it is only manifested if the last bath or solution through which the silk has passed contained an acid salt or free acid. Silk which has been worked in a neutral or alkaline bath—for example, a soap bath—possesses no "scroop." No sufficient explanation of the action of acid in producing scroop has been given, but it is not improbable that acids cause the fibres to become more rough or irregular on the surface, so that when submitted to pressure they slip past each other with a jerky movement.

When it is desired to impart scroop to the silk after dyeing, it is submitted to a special treatment. Very often it is first passed through a weak soap bath or an oil emulsion, and then into a weak acid bath, or it is introduced into a bath which is both oily and acid.

In order to develop all the qualities of softness and brilliancy of which silk is capable, it is submitted (while still in the form of hanks of yarn) to the following mechanical operations :—

Shaking Out Silk (Fr. *sécouage*).—The object of this is to open out or beat out the hanks of silk, and to give the latter a uniform appearance by removing all tendency to curl or wrinkle. It is generally performed after drying, but if before, the drying is greatly facilitated. It consists in hanging the hank of yarn on a strong smooth wooden peg fixed to the wall, and inserting a smooth wooden rod in the loop, which is then vigorously and quickly pulled. The point of suspension is frequently changed, and the shaking out is repeated. The operation may also be done by a machine of M. César Corron, St. Etienne, with the utmost regularity.

Stringing or Glossing Silk (Fr. *chevillage*).—This operation, which was originally only performed in conjunction with the "shaking-out" for the purpose of straightening the threads, and dressing the hanks after diverse operations of the dye-house, has now acquired increased importance, particularly in the case of souples. With these it forms the final operation, the silk being

56 TEXTILE FABRICS.

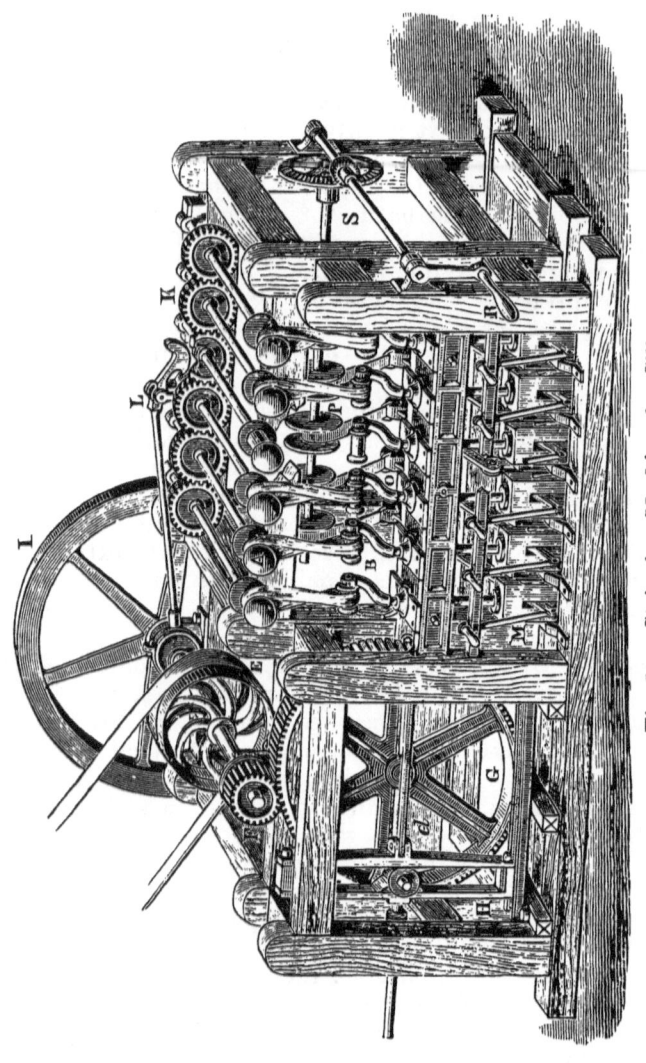

Fig. 15.—Stringing Machine for Silk.

operated upon in the dry state. Its object in this case is to complete the separation of the double silk fibre into its constituent fibres, and to add lustre.

The operation consists in twisting the hanks of silk

when perfectly dry. They are hung on pegs, as in the last operation, which it generally succeeds. A stated and progressive tension is thus given, which adds softness and brilliancy to the fibres. This operation can also be performed with a machine, a representation of which is given in Figs. 15 and 16.

The stringing machine (Fig. 15) is composed of a series of horizontal pegs A, which can be made to revolve by means of the lever and ratchet L and the cog-wheels K. A second series of horizontal rollers B, situated directly beneath the pegs A, are fixed on elbow-shaped spindles. They are capable of two movements, namely, that of revolving on their own axis—*i.e.* the horizontal axis of the

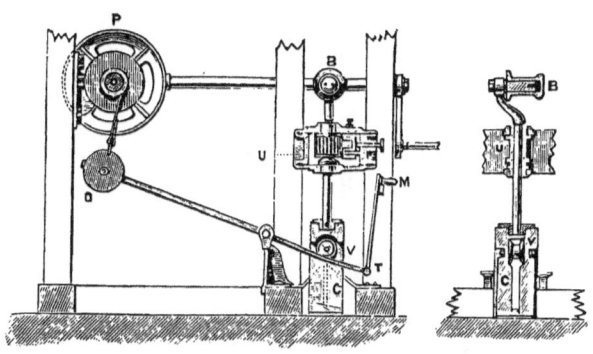

Fig. 16.—Details of Silk-stringing Machine.

elbow—as loose pulleys, and also at right angles to this, by the revolution of the vertical spindle of the elbow. This last movement is caused as follows: Each spindle is supported by a collar on a box U, enclosing a toothed-wheel, which is revolved by a ratchet; this is actuated by the backward and forward movement of the horizontal arm *d* imparted by the frame H, through the medium of the toothed wheels F and G and the pulley E.

Each spindle is also capable of sliding up and down. Attached to the bottom of the roller-pegs B (Fig. 16) are the weights C, which can either be raised separately by the jointed levers M pressing against the pulley V, or all together by means of the handle R. When this handle is turned, the toothed wheels s (Fig. 15) cause the pulley P to revolve, the counterpoise O (Fig. 16) descends, and the weights C, together with spindles and roller pegs B, are raised.

Suppose the hanks of silk are slung over the pegs A and B; by the action of the ratchet and lever D the roller pegs B revolve and the hanks are twisted, the pegs, weights, etc., being at the same time raised by the shortening of the hanks. Automatically the movement of the lever d is reversed, the hanks untwist, and are kept in the stretched condition by the descending weights. At a given moment, namely, just when the hanks are entirely untwisted, the lever L comes into play, and causes a slight rotation of the pegs A, so that the hanks become suspended in a fresh position, to be again twisted by the action of the lever d. The whole of the movements are automatic, and by a few repetitions of this twisting, untwisting, and displacement of the hanks, the operation is complete.

In some cases (as with sewing silks), stringing consists simply in twisting the hanks of silk as tightly as possible, then placing the peg in a locked position, and leaving the hanks in this tightly twisted condition for several hours. The operation may be repeated frequently during from 10-15 days. Its object is to give increased lustre.

Lustring Silk.—This operation, effected by means of the machine represented in Fig. 17, serves to impart the maximum of brilliancy to the fibre. It also facilitates the subsequent winding. The dyed and dried, or sometimes incompletely dried, silk is submitted to a gentle stretching between two polished steel rollers, C and D, revolving in the same direction, and enclosed in a cast-iron box, the lid A and side B of which can be rapidly removed when necessary. During the rotation of the cylinders, steam at a moderate pressure is allowed to enter. The stretching is effected by drawing the roller C away from D by means of the hook F, actuated by the cog-wheels at E.

The brilliant mother-of-pearl lustre possessed by silk undoubtedly gives it the place of honour among all textile fibres.

Tenacity and Elasticity of Silk.—The specific gravity of silk is 1·367. Its tenacity and elasticity are remarkably great. The former is said to be little inferior to that of a good quality of iron wire of equal diameter, while the latter is such that a silk fibre can be stretched 0·14-0·2 of its original length without breaking. These properties are taken advantage of in the operations of shaking out, stringing, and lustring, just described. The finest silk is proportionately the strongest and most tenacious. Damp silk is less tenacious and more elastic than dry silk.

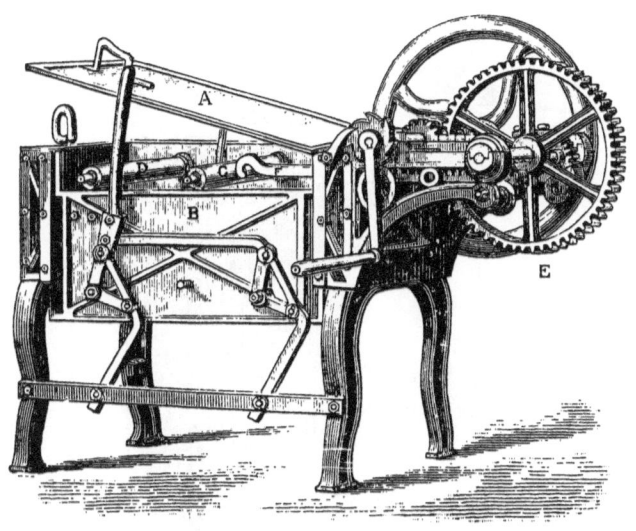

Fig. 17.—Silk-lustring Machine.

If perfectly dried silk is wetted with water, it contracts about 0·7 per cent., and still more if the water contains mineral or organic substances which penetrate the fibre and cause it to swell up. These effects take place during the various operations of dyeing; hence the necessity of stringing, stretching, and lustring, above alluded to, in order to prevent or counteract the contraction.

The tenacity and elasticity of raw-silk reside largely in its external coating of silk-glue. By boiling-off with soap, it loses 30 % of its tenacity, and 45 % of its elasticity.

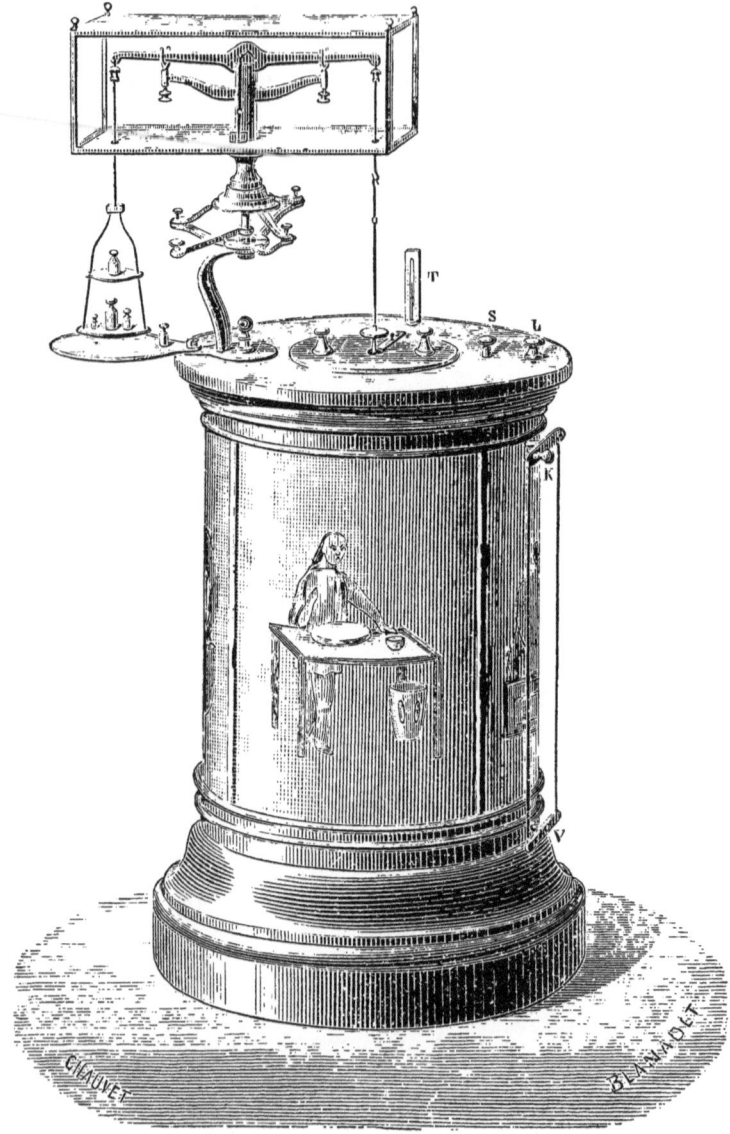

Fig. 18.—Conditioning Apparatus.

These properties vary in weighted silk, according to the nature of the weighing. If the fibre is simply coated with such substances as gelatin, albumen, starch, etc., the tenacity will be as a rule increased, but if the weighting materials employed penetrate the substance of the fibre, and cause it to swell in a greater or less degree, the natural properties of the silk will be modified accordingly. Some agents, like the simple colouring matters, have no appreciable influence, while others, as astringents and metallic salts, when used in large excess, gradually destroy the valuable properties of silk entirely.

If silk is heated to 110° C. it loses all its natural moisture, but remains otherwise quite unchanged. Exposed to 170° C., and higher, it soon begins to decompose and carbonise. If a silk fibre be inserted in a flame it has the appearance of fusing like wool, but it does not give off quite such a disagreeable odour.

Silk is a very bad conductor of electricity, and since it readily becomes electric by friction, this condition, once acquired, is very persistent, and is apt to become a source of trouble during the mechanical operations involved in manufacturing. The most effective mode of overcoming the difficulty is to keep the atmosphere of the workrooms in a suitable state of humidity.

In its boiled-off and pure state silk resists ordinary decay most thoroughly, and it is rarely attacked by insects.

Conditioning Silk.—If raw silk be kept in a humid atmosphere it is capable of absorbing 30 % of its weight of moisture without this being at all perceptible. This circumstance, coupled with the high price of raw silk, makes it of very great importance to those who trade in it to know exactly what weight of normal silk there is in any given lot which may be the subject of commercial dealings. To ascertain this information there have been established, in the principal centres of the textile industry, both in this country and on the Continent, so-called conditioning establishments, as in Lyons, Crefeld, Zurich, Bâle, Turin, Milan, Vienna, Paris, London, Bradford, Manchester, etc. etc.

Fig. 18 shows the external appearance of the essential apparatus of such an establishment, namely, the desiccator. It consists of an enamelled cylindrical hot-air chamber. One arm of a fine balance sustains a crown of hooks, to which are attached the skeins of silk to be dried. The suspending wire passes through a small opening in the cover of the cylinder. The other arm of the balance carries the ordinary pan for weights.

Fig. 19 gives a vertical section of the chamber. Hot air at 110° C. enters by the tube A from a stove situated in a cellar below, passes into the space B, and thence by thirty-two vertical tubes, t, placed between the two concentric cylinders C and D, it enters the upper portion of the inner cylinder D. The hot air descends, dries the silk, and escapes by the tubes E, which communicate with the exit flue. The apparatus is provided with a valve V, actuated by the lever K (Fig. 18) for regulating or shutting off the current of hot air.

The air which passes outside the brickwork of the stove, and is thus heated only to a moderate degree, passes upwards between the cylinders C and D into the space r; by means of the button L, which actuates a slide-valve, its entrance into the central chamber can be regulated. By means, therefore, of the lever K and the button L, the supply of hot and cold air into the central chamber can be regulated to a nicety, and the temperature of the mixture is ascertained by the thermometer T. The button S actuates the valve M, which cuts off communciation with the exit flue and stops the current of air during a final weighing operation.

Several hanks of silk are taken from the bale to be tested and divided into three lots, in order to be able to make two parallel determinations, and a third if necessary. The weight is first rapidly taken, under ordinary circumstances, on a fine balance; the hanks are then suspended in the desiccator and counterpoised, and the hot air current is allowed to circulate till no further loss of weight takes place. One operation may last from $\frac{1}{2}$-$\frac{3}{4}$ hour.

The average loss of weight usually met with is about 12 %. Absolutely dry silk is not reckoned as the standard article, but such as contains about 90 % dry silk and 10 % moisture. The legal weight is really obtained by adding 11 % to the dry weight.

Chemical Composition of Silk.—The silk fibre has been

the subject of numerous chemical researches, the general result of which may be summed up by saying that it is composed essentially of two distinct parts: first, that constituting the central portion of the fibre, and secondly, a coating, or envelope, consisting apparently of a mixture of substances mostly removable by hot water, or, at

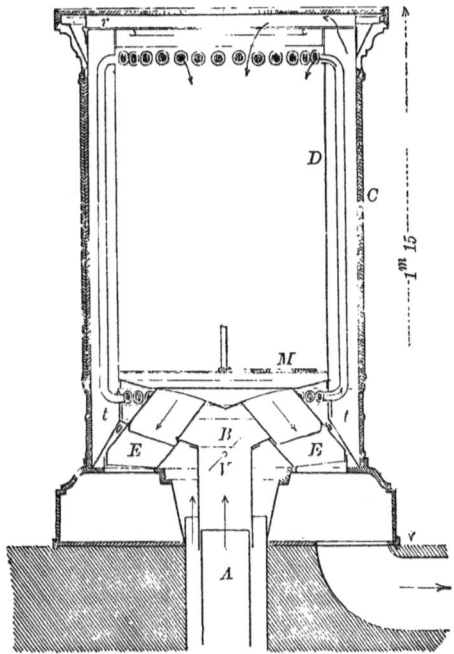

Fig. 19.—Section of Conditioning Chamber.

any rate, by solvents which have little or no action on the central portion.

In order to determine the character and amount of these several substances, Mulder submitted raw Italian silk to the successive action of boiling water, alcohol, ether, and hot acetic acid, and in this way obtained in a comparatively pure state the central silk substance, to which he assigned the name *Fibroïn*. The following numbers give the results of his analysis :—

	Yellow Italian silk.	White Levant silk.
Silk fibre (fibroïn)	53·35	54·05
Matters soluble in water	28·86	28·10
alcohol	1·48	1·30
ether	0·01	0·05
acetic acid	16·30	16·50
	100·00	100·00

By a further examination of the substances which each solvent had extracted, he arrived at the following more detailed analysis:—

	Yellow Italian silk.	White Levant silk.
Fibroïn	53·37	54·04
Gelatin	20·66	19·08
Albumen	24·43	25·47
Wax	1·39	1·11
Colouring matter	0·05	0·00
Resinous and fatty matter	0·10	0·30
	100·00	100·00

Mulder explains that on evaporating the aqueous solution to dryness, the residue would not entirely redissolve in water. This insoluble portion, therefore, and that which is dissolved by acetic acid, has been reckoned as albumen. Exception has been taken by Bolley with regard to the presence of this albumen, for if it is borne in mind that the temperature employed in killing the cocoons, and that of the water used during the reeling process, is such as to coagulate any albumen which might possibly be present, it is highly improbable that raw silk would contain any soluble albumen.

Further, it has been shown that living cocoons, which have not therefore been submitted to any steaming process, but have simply been opened and heated with tepid water, contain *no* albumen. Fibroïn itself, too, is known to be somewhat soluble in strong acetic acid, so that it may, on the whole, be concluded that what Mulder found to be soluble in acetic acid was not albumen, but altered fibroïn, and that the percentage of this latter substance in silk which he gives, is too low.

By heating raw silk for several hours with water at 133° C., a residue of fibroïn is obtained which, after the removal of fatty matter by ether, and colouring matter by alcohol, represents 66 % of the weight of the silk. Even this figure may possibly be too low, since the usual loss in practice during the operation of "boiling-off" is 25-30 %.

The percentage composition of pure *Fibroïn* has been variously stated, probably owing to the different hygrometric states of the fibres examined. Cramer gives the formula as $C_{15}H_{23}N_5O_6$.

Sericin is that portion of silk which is soluble in warm water and can be precipitated from its solution by lead acetate. By submitting this precipitate to a somewhat tedious series of operations—such as washing with water, suspending in water and decomposing with sulphuretted hydrogen, filtering, and evaporating the filtered solution, precipitating and extracting with alcohol, then with ether —it is possible to obtain the essential constituent of the external envelope of the silk fibre as a colourless, odourless, tasteless powder. It swells up in cold water, and is somewhat more soluble in hot water than gelatin. A 6 % solution of it gelatinises on cooling, and its solutions are precipitated by alcohol, tannic acid, and metallic salt solutions. Altogether, its physical and chemical properties are very similar to those of glue and gelatin; hence sericin is often called silk-glue, sometimes also silk-gum. Its chemical composition is represented by the formula— $C_{15}H_{25}N_5O_8$.

It is distinct from ordinary glue, however, according to some observers, since when boiled with dilute mineral acids it yields different products.

If the formulæ given for fibroïn and sericin be compared, a relationship is apparent which may be expressed by the following equation :—

$$\underset{\text{Fibroïn.}}{C_{15}H_{23}N_5O_6} + O + H_2O = \underset{\text{Serici}}{C_{15}H_{25}N_5O_8}$$

Although these formulæ can only be considered as representing approximately the percentage composition of these two bodies, the above comparison has been taken by some as an indication that originally—*i.e.* at the moment of secretion—the silk fibre is probably a homogeneous substance, which, by the action of the air and moisture, rapidly becomes altered superficially. This view is supported by the observation that if moist fibroïn be left exposed to the air for a lengthened period it becomes partially soluble in water. Bolley and Rosa have found also that the silk-bags taken from living worms are composed almost entirely of fibroïn, since only 1·7 % is soluble in boiling water, and the elementary analysis is consistent with the formula of fibroïn. The physiological studies of Duseigneur, and especially his examination of the transverse section of the silk-bag, already alluded to, appear to contradict this view.

Influence of Reagents on Silk.—In contact with various liquids silk not only absorbs them rapidly, on account of its great porosity, but sometimes retains them with extreme tenacity; this is the case, *e.g.* with alcohol and acetic acid. For the same reason it has great aptitude for fixing mordants and colouring matters.

Action of Water on Silk.—Prolonged boiling with water removes from raw silk its silk-glue, but it has little effect upon the fatty, waxy, and colouring matters present. The tenacity of the fibre is reduced even more than by the ordinary methods of ungumming with soap solutions. A similar solvent action is exercised by all liquids; for this reason it is not customary to mordant silk with hot solutions, and the dyeing is conducted at a temperature as low as circumstances will permit.

Action of Acids on Silk.—Speaking generally, concentrated mineral acids rapidly destroy silk, but if sufficiently diluted their action is insensible. Warm dilute acids, however, dissolve the sericin of raw silk, and hence these may be used in ungumming (soupling). Concentrated sulphuric acid dissolves silk, giving a viscous brown liquid; on diluting the latter with water a clear solution is obtained, from which the fibroïn is precipitated on the addition of tannic acid.

Concentrated nitric acid also rapidly destroys silk; but if diluted, the latter is only slightly attacked and coloured yellow, in consequence of the formation of xanthoproteïc acid. This reaction is made use of in distinguishing silk from vegetable fibres. Formerly it was also utilised in dyeing, but the method is not to be commended, since the colour is produced at the expense of the silk itself, which must inevitably be weakened by the process.

Hydrochloric acid, if applied in the gaseous state, destroys the fibre without liquefying it, but a concentrated aqueous solution readily dissolves it. Hydrochloric acid 38° Tw. (Sp. Gr. 1·19) when applied cold, dissolves an equal weight of silk without even then being saturated. Dilute hydrochloric acid has no sensible action, except upon the sericin of raw silk, which it more or less removes.

Phosphoric and arsenic acids in dilute (5 %) aqueous solution act like other weak acids in removing the sericin from raw silk, and have been proposed as ungumming agents instead of soap, but they are not used in practice.

Permanganic acid, either in the free state or in combination with potassium, acts energetically on silk; it oxidises and colours the fibre brown by deposition of hydrated manganic oxide. If this be removed by immersion in a solution of sulphurous acid, the silk is left in a remarkably pure white condition. Although recommended on this account for bleaching silk, it is not altogether suitable, since the silk thus bleached always has a tendency to become yellowish under the influence of alkalis.

Sulphurous acid is used in bleaching silk.

Chromic acid and chromates, like permanganic acid, oxidise silk, leaving the fibre of a pale olive tint.

The action of organic acids on silk has been little studied; it varies, no doubt considerably, according to their concentration, temperature, etc.

Hot dilute organic acids remove the sericin from raw silk but do not affect the fibroïn much. Cold glacial acetic acid removes the colouring matter from yellow raw-silk without dissolving the sericin. Silk is entirely dissolved when heated under pressure with acetic acid.

Action of Alkalis on Silk.—Concentrated solutions of caustic soda and potash rapidly dissolve raw-silk, especially is applied warm.

Caustic alkalis, sufficiently diluted so as not to act appreciably upon the fibroïn, will dissolve off the sericin, and have been tried as ungumming agents. For ordinary use, however, they must be avoided, since the silk is always left impaired in whiteness and brilliancy.

Pure ammonia solution, even if used warm, has no sensible action on boiled-off silk, but if it is at all impure the silk becomes dull and dirty from absorbed tarry matter. Ammonia seems to favour the absorption by silk of salts of calcium, magnesium, etc.

Alkaline carbonates act like the caustic alkalis, but in a less energetic manner, and they are not employed as ungumming agents. Of all alkaline solutions, those of soap have the least injurious effect. When used hot, they readily remove the sericin from raw silk, and leave the fibroïn lustrous and brilliant; hence soap is *par excellence* the ungumming agent employed. Borax acts somewhat like soap, but cannot replace it in practice.

If raw silk be steeped for twenty-four hours in clear, cold lime-water, it swells up considerably, the lime seeming to have a strong softening action on the sericin; when this is removed by dilute acid and a subsequent soap bath, the fibroïn seems not to have suffered otherwise than by the loss of its natural brilliancy. Prolonged contact, however, with lime-water renders silk brittle and disorganised.

Chlorine and hypochlorites attack and destroy silk rapidly, and cannot be used as bleaching agents. Applied in weak solutions, with subsequent exposure of the fibre to the air, they cause the silk to have an increased attraction for certain colouring matters.

Action of Metallic Salts on Silk.—If silk is steeped in cold solutions of several metallic salts, *e.g.* of lead, tin, copper, iron, aluminium, etc., it absorbs and even partly decomposes them, so that less soluble basic salts remain in union with the fibre. The methods of mordanting silk with aluminium, tin, and iron salts depend upon this fact. Sometimes, as in the case of ferric and stannic salts, the quantity of basic salt which may be precipitated on the fibre is sufficient to serve as weighting material.

Concentrated zinc chloride, 138° Tw. (Sp. Gr. 1·69), made neutral or basic by boiling with excess of zinc oxide, dissolves silk, slowly if cold, but very rapidly if heated, to a thick gummy liquid. This reagent may serve to separate or distinguish silk from wool and the vegetable fibres, since these are not affected by it. If water be added to the zinc chloride solution of silk, the latter is thrown down as a flocculent precipitate. If this is washed free from zinc salt and dissolved in ammonia, it is said that the solution may serve to cover cotton and other vegetable fibres with a coating of silk substance. Dried at 110°-115°

C., the precipitate acquires a vitreous aspect, and is no longer soluble in ammonia.

An ammoniacal solution of cupric hydrate dissolves silk, the solution not being precipitated by neutral salts, sugar, or gum, as is the case with the analogous solution of cotton. An ammoniacal solution of nickel hydrate also dissolves silk.

A most excellent solvent for silk is an alkaline solution of copper and glycerine, made up as follows: dissolve 16 g. copper sulphate in 140-160 c.c. distilled water, and add 8-10 g. pure glycerine (Sp. Gr. 1·24); a solution of caustic soda is dropped gradually into the mixture till the precipitate at first formed just re-dissolves; excess of NaOH must be avoided. This solution does not dissolve either wool or the vegetable fibres, and may serve, therefore, as a distinguishing test.

Action of Colouring Matters on Silk.—Generally speaking, silk has a very great affinity for the monogenetic colouring matters. It can be dyed direct with the aniline colours, for example, with the greatest facility. It has, however, little attraction for mineral colouring matters.

An examination of sections of dyed silk reveals the fact that the colouring matter (or the mordant) penetrates the substance of the silk fibre to a greater or less degree, according to the solubility of the colouring matter, the duration of the dyeing process, and the temperature employed. If the silk is dyed only for a short time, a section of the fibre shows an external concentric zone of colour, while if the dyeing operation is continued sufficiently long, it is coloured right to the centre. If a mixture of two colouring matters be applied, either simultaneously or successively, both are absorbed, the more soluble, or that which has been allowed to act longest, penetrating the fibre most deeply. Externally a mixed effect is produced in this case, but a section of the fibre reveals in most cases two concentric zones of colour. Silk thoroughly mordanted with a ferric salt presents in section a uniform yellow tint; if dyed subsequently in an acidified solution of potassium ferrocyanide, the ferric oxide deposited in the silk gives place to Prussian blue, at first in the outer portions only, but by degrees even in the centre, especially if the temperature of the bath be raised. A similar effect is produced if a bath of tannin be substituted for that of potassium ferrocyanide. It is indeed difficult to say what number of substances might be successively absorbed by the silk, and penetrate it either by juxtaposition or by reacting upon each other.

The action of colouring matters on raw silk is similar; but in many cases, as in the black dyeing of souples, the colouring matter is situated principally in the external silk-glue which, becoming brittle through the large amount of foreign matter it then contains, breaks up and assumes the form of microscopic beads.

CHAPTER V.

COTTON BLEACHING.

Object of Bleaching.—Raw cotton is contaminated with several natural impurities, and although these are comparatively small in amount, they impair the brilliancy of the white belonging to pure cellulose. Hence cotton yarn as it leaves the spinner has invariably a soiled or greyish colour. When such yarn is woven it is still further contaminated with all the substances (amounting sometimes to 30 %) which are introduced during the sizing of the warps, as china clay, grease, etc.

Bleaching consists in the complete decolorising or removal of all these natural and artificial impurities, either for the purpose of selling the goods in the white state, or in order to make them suitable for being dyed light, delicate, and brilliant colours.

Bleaching Raw Cotton.—Although raw cotton is now largely dyed, it is seldom bleached in this form, because it becomes more or less matted together. As a rule, the only treatment previous to dyeing which it receives is that of boiling with water until thoroughly wetted.

Bleaching Cotton Yarn.—When cotton yarn has to be dyed black or dark colours, it is, as a rule, not bleached, but merely boiled with water till thoroughly softened and wetted. For light colours the dyer frequently effects a rapid, though perhaps more or less incomplete, bleaching by passing the wetted yarn through a boiling weak solution of soda-ash, then steeping it for a few hours in a cold weak solution of chloride of lime or hypochlorite of soda. It is then washed in water, steeped in dilute hydrochloric acid, and finally well washed.

A more complete and thorough bleaching is that effected by the operations now to be briefly described.

"Warps" are loosely chained by hand or machine, in order to reduce their length. If the yarn is in hanks it is either retained in that form, or linked together to form a chain, the latter being the better and more economical method.

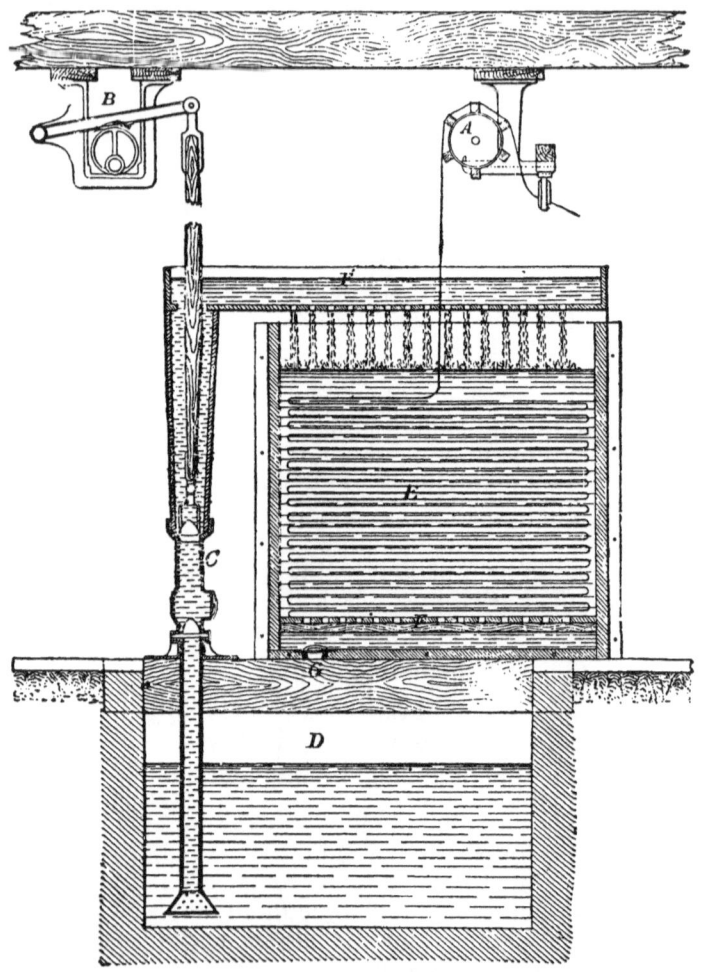

Fig. 20.—Apparatus for Chemicking, Souring, and Washing.

1. *Ley boil.*—For 1,500 kg. yarn, boil six hours with 2,000 litres water and 300 litres caustic soda 32° Tw. (Sp. Gr. 1·16); steep in water for forty-five minutes and wash.

COTTON BLEACHING.

2. *Chemicking.*—Steep the yarn for two hours under sieve in a solution of bleaching powder 2° Tw. (Sp. Gr. 1·01), then wash for half an hour under sieve.

3. *Souring.*—Steep the yarn for half an hour under sieve in dilute sulphuric acid 1° Tw. (Sp. Gr. 1·005), then wash for half an hour under sieve and afterwards through washing machine.

If the yarn is intended to remain white and not to be dyed, it is run through a so-called "bluing" machine with hot soap solution and blue (ultramarine, etc.), then hydro-extracted and dried.

When bleaching cotton thread, owing to its closer texture, the first three operations are repeated

The boiling (also called "bowking" or "bucking") with caustic soda solutions takes place in large iron boilers or "kiers." These are either open or provided with a lid capable of being screwed down, in order to be able to boil with a slight pressure of steam.

The usual order of procedure is first to fill the kier with the yarn, and after blowing steam through for an hour or so, to run in the soda solution and boil for 10-12 hours.

The operations of chemicking, souring and washing under sieve, are carried out by means of the arrangement shown in Fig. 20. It consists of a stone tank E, with a false bottom F, and a valve G, communicating with the cistern D below; B is the shaft which works the pump C; F' is a movable perforated drainer or sieve covering the whole surface of the tank E; A is a winch for drawing the chain of yarn into the tank. Supposing the tank to be packed with yarn, the pump is set in motion, the liquid in D is thus raised to the sieve F', whence it showers down on the yarn below. It filters, more or less rapidly, through the yarn and collects again in the tank D to circulate as before.

Complete separate arrangements of this kind are required both for chemicking and for souring, but the washing under sieve is performed in either set of tanks as required, it being only necessary to stop the pump, close the valve G, and allow water to flow from a tap placed over the sieve, and to escape at the bottom of the tank E by a separate plug-hole into the nearest drain. The final washing after souring is best given by means of a square beater washing machine.

The "bluing" machine referred to is essentially the same in construction as the final washing machine, the main difference being that the square beater is replaced by a round roller, and that the upper squeezing roller is covered with cotton rope and rests loosely with its own weight on the lower one. As the cotton yarn, soaked with soap solution and blue, passes rapidly between the squeezing rollers, the irregularities produced by the plaiting or linking impart a constant jumping motion to the upper roller, and the liquid is effectually beaten and pressed into the heart of the yarn, thus enhancing considerably the purity of the white.

Bleaching Cotton Cloth or Calico.—The mode of bleaching is varied according to the immediate object for which the bleached calico is intended; thus, one may distinguish between the Madder-bleach, the Turkey-red-bleach, and the Market-bleach.

Madder-bleach is the most thorough kind of calico-bleaching, and was originally so-called because it was found specially requisite for those goods which had to be printed and subsequently dyed with madder. Its object is to effect the most complete removal possible of every impurity which can attract colouring matter in the dye-bath, so that the printed pattern may ultimately stand out in clear and bold relief on a white background of unstained purity. Although the madder-bleach is in general use among calico-printers, it may be also adopted by dyers whenever the calico is to be dyed subsequently in light and delicate colours, or if absolute freedom is desired, from any impurity which resists the fixing of the colouring matter to be applied.

Stamping and Stitching.—For the purpose of subsequent recognition, the ends of each piece are marked with letters and figures, by stamping them with gas-tar or other substance capable of resisting the bleaching process. The pieces are then stitched together, end to end, by machinery.

Singeing.—This operation consists in burning off the nap or loose fibres which project from the surface of the cloth, since these interfere with the production of fine

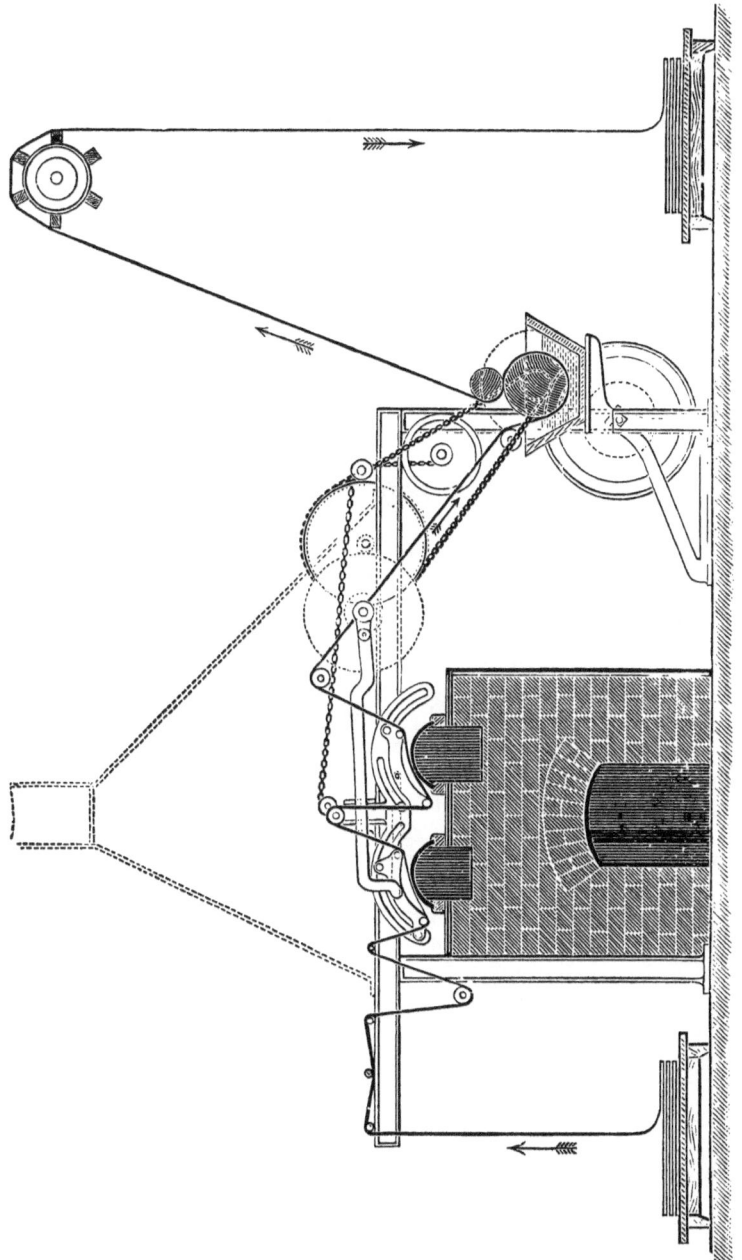

Fig. 21.—Plate Singeing Machine by Mather and Platt.

impressions during the printing process. It is performed by rapidly passing the cloth in the open width over red-hot plates or cylinders, or over a row of gas flames.

Fig. 21 illustrates the Mather and Platt plate singeing machine. By means of a furnace, two copper plates are kept at a red heat, and the piece is passed rapidly over them in the direction shown by the arrows. Immediately after passing the plates the piece passes down into a trough of water so that any adhering sparks are at once extinguished. After this the piece is squeezed and "fliped down" in the usual manner. The chief difficulty with these machines is to prevent the plates rapidly cooling at the place where the piece touches them. In the machine illustrated this is avoided by traversing the bars which press the cloth down on to the plate, so that the piece is continually changing its position whilst every portion of the surface of the plate is utilised in turn. In another machine the same end is gained in a somewhat clumsier manner by using tubular plates through which the flames from the furnace pass, the plates themselves being continually rotated. The amount of singe is regulated by raising or lowering the bars guiding the piece, which is done by means of the chain shown. There is also a sheet metal hood which covers the greater portion of the machine and is provided with a flue through which the products of combustion are led away.

This method of singeing is, as a rule, only now used for thick or heavy cloth, and for light goods is entirely superseded by gas singeing, a typical machine for which work is shown in Fig. 22.

In this, the Mather and Platt machine, the gas flame bears directly against the cloth, and is not used to heat bars as in some cases. The cloth is fed from the right of the illustration, and, after passing tensioning bars, is passed across a slot in the underside of an exhausting chamber connected with a fan, so that the flame from the burner immediately below is drawn through the cloth by the current of air. The burner and the above-mentioned exhausting chamber are shown on an enlarged scale at the right of Fig. 22, in which the small arrows show the passage of the gas and flame, whilst a larger arrow points to a water jacket which surrounds the exhausting cham-

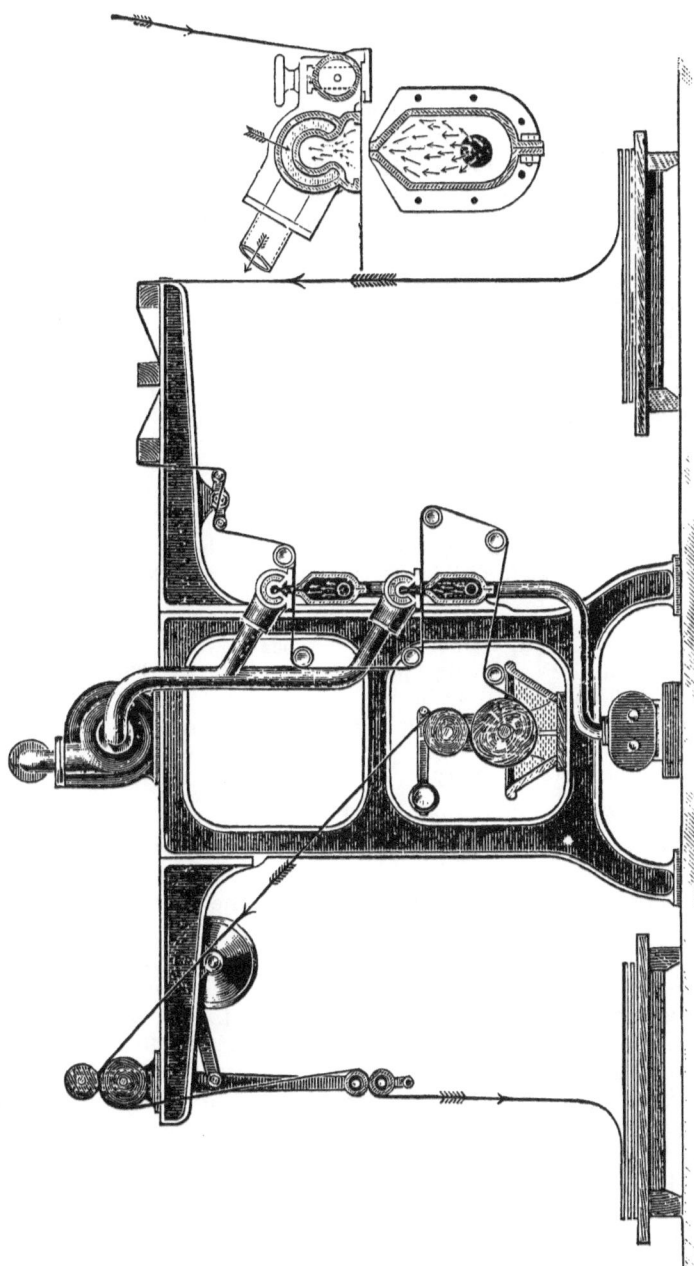

Fig. 22.—Gas Singeing Machine, by Mather and Platt.

ber, and through which a current of cold water is maintained. As a rule, one burner working on this principle is enough to singe light fabrics, but in case it is necessary to thoroughly singe both sides, two burners are used, as shown in the illustration. The cloth passes over one, is turned by passing round two guide rods, and then presents its other face to a second burner. It then passes through the usual trough of water and out of the machine in the direction marked by the arrows.

The preliminary work of stamping, stitching, and singeing is succeeded by the bleaching operations proper, which, for 24,000 kg. = 23·6 tons cloth and with low pressure kiers, may be summarised as follows :—

1. Wash after singeing.
2. Lime-boil: 1,000 kg. (2204·6 lb.) lime, boil 12 hours : wash.
3. Lime-sour: hydrochloric acid, 2° Tw. (Sp. Gr. 1·01); wash.
4. Ley-boils : 1st, 310 kg. (749·4 lb.) soda ash, boil 3 hours.

    2nd, 860 kg. (1896 lb.) soda ash, 380 kg. resin, 190 kg. (418·9 lb.) solid caustic soda, boil 12 hours.

    3rd, 380 kg. (837·74 lb.) soda ash, boil 3 hours ; wash.
5. Chemicking : bleaching powder solution, $\frac{1}{4}°-\frac{1}{2}°$ Tw. (Sp. Gr. 1·00125—1·0025) wash.
6. White-sour : hydrochloric acid, 2° Tw. (Sp. Gr. 1·01), pile 1—3 hours.

    Wash, squeeze and dry.

1. *Wash after Singeing.*—The object of this operation is to wet out the cloth and make it more absorbent, also to remove some of the weaver's dressing. This was formerly effected by simply steeping the cloth in water for several days until, by the fermentation induced, the starchy matters were rendered more or less soluble. At present, printers' calicoes are not, as a rule, heavily sized, and a simple wash is sufficient. The pieces are drawn direct from the adjacent singeing house, guided by means of white glazed earthenware rings (" pot-eyes "), through the washing machine ; they are at once plaited or folded down on the floor and there allowed to lie " in pile " for some hours to soften. By this first operation, frequently called " grey-washing," the pieces, hitherto in the open width, assume the chain form, which in many cases they retain throughout the whole of the succeeding operations

COTTON BLEACHING. 79

2. *Lime-boil* ("Lime-bowk").—The pieces are now run through milk of lime, a portion of which they absorb. They are then passed by overhead winches into a kier, and there plaited down and well packed.

The most modern method is that which is made possible by the Mather and Platt kier shown in Fig. 23. The goods

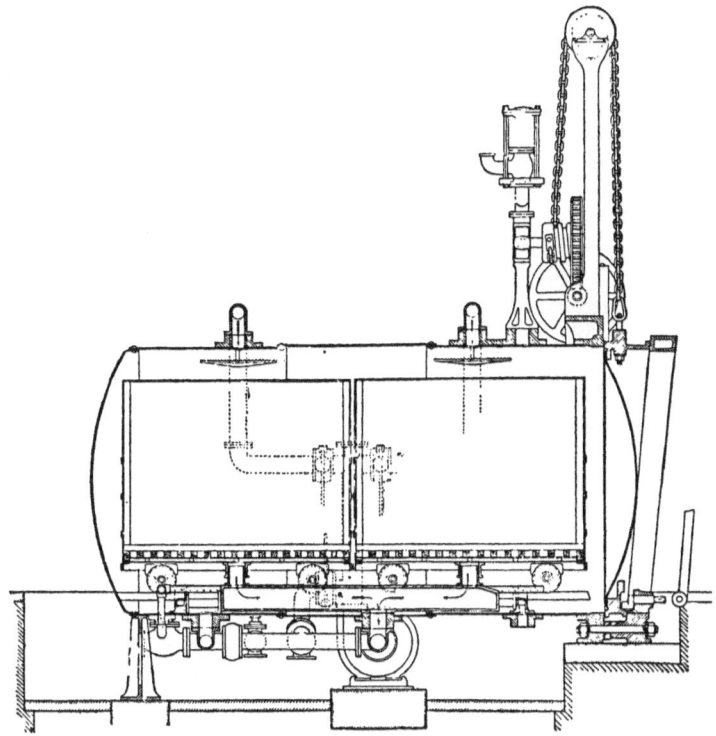

Fig. 23.—Longitudinal Section of the Mather Patent Kier.

are run into trucks, and these can be filled whilst preceding pieces are in the kier. When a change is to be made, the cover of the kier (shown to the right of the illustration) is drawn up by a winch and chain. Rails run right up to the mouth of the kier and are continued inside, so that the trucks can be run in or out without much labour. Each truck has a perforated bottom, and two, as a rule, are in the kier at a time. When the kier is again closed steam is admitted, and then the liquor is run in, falling on two corrugated discs and then in a spray to each truck. It flows down through the material, and is drained off at the bottom, to be again pumped up and delivered above. The arrows show the direction of the liquor to the centrifugal pump below the kier.

In kiers specially designed for cloth bleaching the cloth is treated in open width, and either boiled whilst on a perforated drum or passed backwards and forwards between two drums whilst in the kier.

The essential action of boiling with lime is to decompose the fatty, resinous, and waxy impurities present in the fabric. They are not removed, but remain attached to the fibre as insoluble lime soaps, which can, however, be readily removed by the subsequent processes.

The colouring matter of the fibre is modified, and any alumina present is also attacked.

No doubt, caustic alkalis would also decompose and at once render soluble the fatty impurities, but lime is cheaper, and is said to attack the resinous matters more energetically.

3. *Lime-sour.*—This operation, also called the "grey sour," is immediately preceded by a washing of the pieces as they come out of the lime-boil. It consists in washing the pieces with dilute hydrochloric acid. During this process the insoluble lime-soaps, resulting from the lime-boil, are decomposed and the lime is removed; any other metallic oxides present are also dissolved out, and the brown colouring matter is loosened. Hydrochloric acid is preferred to sulphuric acid because it forms a more soluble compound with the lime. Care must be taken to maintain the strength of the dilute acid as uniform as possible, both by having a regular flow of fresh acid from a stock cistern, and by making occasionally rapid acidimetrical tests. After souring, it is advisable not to leave the pieces long in their acid state, for fear of the exposed portions becoming tender, but to wash them as soon as convenient. It is very essential, too, that this washing be as complete as possible, otherwise a tendering action may take place during the following process.

4. *Ley-* or *Lye-boil.*—The object of this operation is

to remove the fatty matters still remaining on the cloth. The fatty matters having been decomposed during the lime-boil, and the lime having been removed from the lime-soaps by the souring, the fatty acids remaining on the cloth are readily dissolved off by boiling with alkaline solutions. The brown colouring matters are also chiefly removed at this stage. The boiling takes place in exactly the same kind of kiers as those used for the lime-boil. Other kiers besides the above, however, are frequently used for both operations.

In the so-called "vacuum kiers," perfect penetration of the cloth by the liquors is obtained by first pumping out the air from the kier before admitting the liquors.

Fig. 24 gives the section of a modern " injector " kier A, filled with cloth; B B are the steam pipes, C is the injector, and D the circulating pipe; F is the liquor pipe, by which water or other liquid may be admitted; E E the draw-off valve and waste pipe. The kier being suitably filled with cloth and liquor, whenever the steam is turned on, the vacuum produced by its condensation in C withdraws the liquor from the kier and causes it to ascend the pipe D, to be at once showered over the pieces at G. A portion of the liquid may temporarily collect at H, but it soon percolates or is drawn through the cloth to the bottom, again to enter the injector. A continual circulation of the liquid is thus maintained.

If open or low-pressure kiers are used, similar to that referred to in cotton-yarn bleaching, the boiling is continued for 10-12 hours; with the injector kier and steam at 50 lb. pressure, 3-4 hours' boiling may be sufficient.

Some bleachers boil 1-3 hours with soda-ash alone, both before and after the resin boil, using 1-2 kg. soda-ash per 100 kg. calico. The first soda-ash boil, though not absolutely necessary, is advisable, in order to neutralise any traces of acid accidentally left in the cloth from the souring. Another plan to avoid tendering, is to let the goods steep in a weak soda-ash solution for a short time, and then to draw it off again before commencing the boiling operation with " resin-soap." This is termed " sweetening " the goods.

The boiling with resin-soap is a very special feature in the madder-bleach. Experiment has shown that resin-

F

soap removes, better than any other substance which has been tried, certain matters, which would subsequently attract colouring matter in the dye-bath.

The boiling with soda-ash solution after the "resin

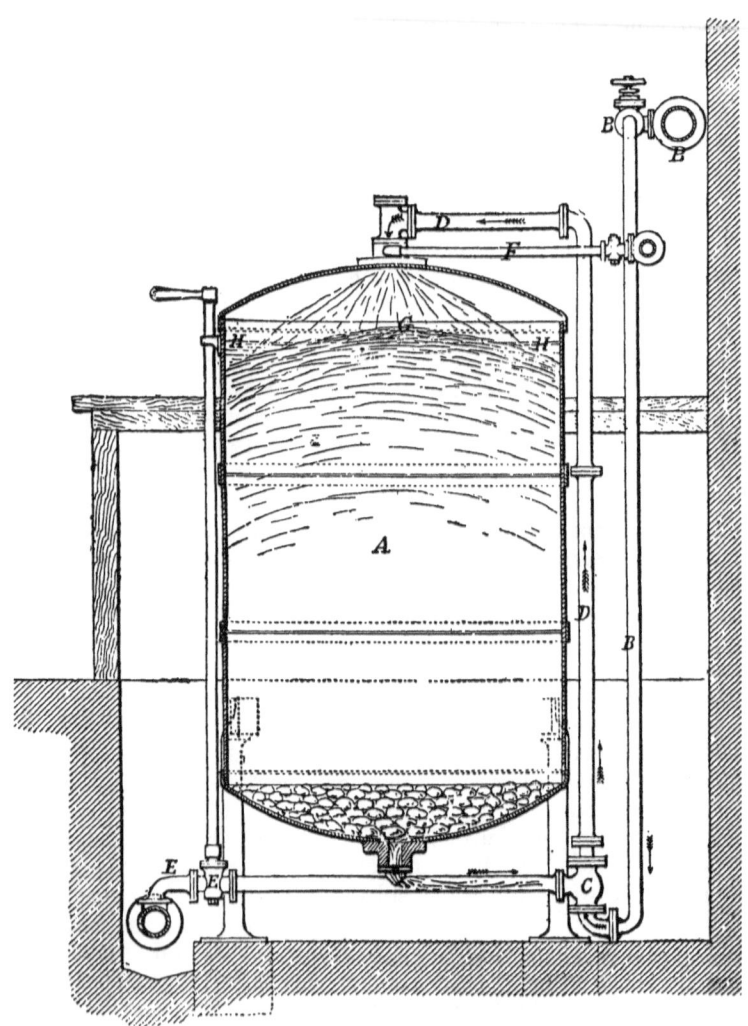

Fig. 24. – Section of Injector Kier.

boil is useful, in order to ensure the complete removal of fatty matters and undissolved resin.

Since the cloth is very liable to contract iron stains if left in the kier too long after the alkaline liquor has been drained away, it is well to wash immediately after the ley-boils.

5. *Chemicking.*—After all the previous operations the cloth still retains a faint yellowish or creamy tint, and the object of this operation is to destroy the traces of colouring matter from which it arises. The pieces are passed through a very dilute solution of chloride of lime or "bleaching-powder," in a "chemicking" machine, which is exactly similar to that employed for washing, and are then allowed, while still moist, to remain in pile and exposed to the air for a few hours or over-night. The bleaching action, which must be considered as one of oxidation, takes place largely during this exposure, hypochlorous acid being then liberated by the action of the carbonic acid of the air.

It is essential that the bleaching-powder solution should not be too strong, otherwise the cloth may be tendered or be partially changed into oxycellulose, and thereby be apt to attract certain colouring matters in the dye-bath, or to contract brown stains during subsequent steaming processes. For the same reason the solution of bleaching-powder should be entirely free from undissolved particles.

The bleaching power of the liquor should be maintained as constant as possible by having a continual flow of fresh bleaching-powder solution into the machine, and by occasionally testing how much of the liquor is required to decolorise a specially prepared standard solution of arsenate of soda, tinted with indigo extract or cochineal decoction.

6. *White-sour.*—This operation does not differ from the lime-sour already described. Its object is to complete the bleaching action by decomposing any "chloride of lime" still in the cloth, also to remove the lime, the oxidised colouring matter, and any traces of iron present. The cloth usually remains saturated with the acid a few hours.

7. The *final washing* must be as thorough as possible, and is usually performed by the square beater machine.

After squeezing, the cloth is again opened out to the full width, previous to drying. This is effected by allowing a lengthened portion of the chain of cloth to hang loosely and horizontally, and in this position to pass between a pair of rapidly revolving double-armed scutchers, which shake out the twists from the horizontal length of cloth. The two arms of the scutcher merely loosen the twists, and whilst in this loose condition the cloth passes, over and under respectively, two spiral rollers. The spiral works outwards from the centre on each roller, so that the piece is subjected towards the selvages to a smoothing-out action. This is generally sufficient to effectually open out the cloth, but for heavy goods one type of Mycock scutcher is provided with spirally wound beater blades in addition to the spiral rollers. By means of a sun and planet motion, the spirally-wound blades receive a rapid rotation in addition to the motion of the scutcher arms which carry them. In this machine there is also a governing action, the piece passing under tension through swivelling bars. When the piece leaves the centre, the overlapping side swivels the bars at its own side, and in that manner changes the angle at which it is delivered to the rolls. That angle throws it nearer the centre of the rolls, and so has a contrary effect, which nullifies the first deviation. Instead of the governor bars being swivelled at their centre, in a more modern Mycock machine a parallel motion is provided at each end of the bars, which prevents the swinging action, to which centre swivels are subject, being set up.

At this stage the piece can often be dried, and is ready for that operation if it is wide enough. Nowadays, however, manufacturers so cut their prices that with plain goods they can afford no more width than necessary in tne piece, and the result is that not enough is allowed for shrinkage. The finisher must make up for this, so that between the scutcher and the drying machine an expander is frequently employed. For light stretches spirally-wound rollers are sufficient if set closely to the nip of the rolls of the drying machine, while two conical rollers and other devices are sometimes used. To get a powerful stretch, however, and bring the cloth out to near loom width, bar expanders are necessary.

The most modern of these is shown in Fig. 25, this being the Mycock five-bar expander. As a rule, a three-bar expander is all that is necessary, and the five-bar is only

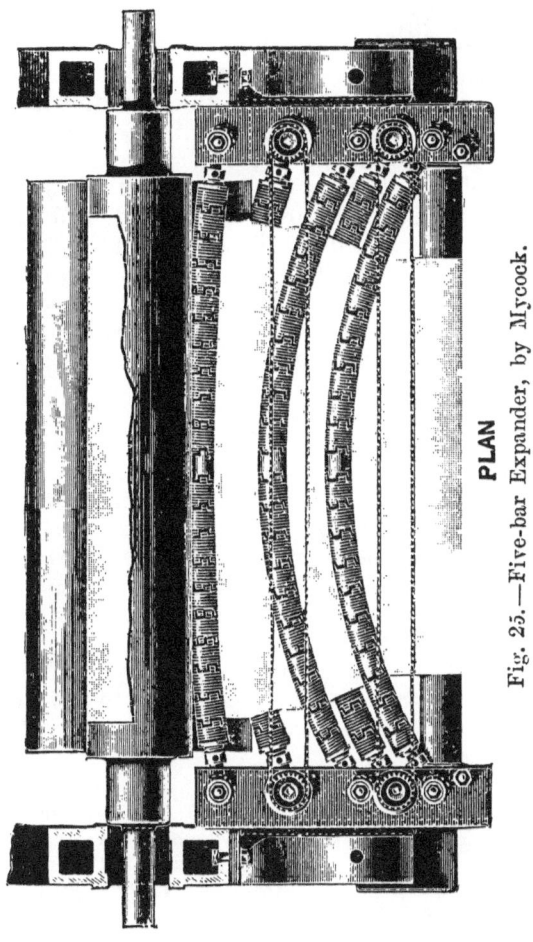

Fig. 25.—Five-bar Expander, by Mycock.

used in the water mangle so as to take the stretched piece right up to the nip of the rolls. This latter is done by making the two extra bars take the curve of a circle of greater diameter, and so get nearer a straight line. These bars are made up of small shell sections, mounted on a central core independently of each other, but fitting into each other so that they all rotate together with the cloth. The outer circumference of the sections are fluted so as to better grip the cloth, but, being mounted on a curved bar, must open out on the convex side to meet the difference between the inner and outer sides of the curve. In thus opening out, the cloth is taken with them and the necessary expansion in width obtained, as can be seen by the course of the cloth in Fig. 25. Regulation of width is made possible by altering the position of the middle bar in the three-bar expander and the second and fourth bars in the five-bar expander. These regulating bars can be raised or lowered at will by the chains and sprocket wheels shown and screw gearing not shown, their height determining the amount of bearing surface which the cloth has on the bars, and therefore the amount of expansion it receives.

The average length of time required for the madder-bleach is 4-5 days.

Turkey-red-bleach is used when calico is intended to be dyed Turkey-red, as then it is not necessary to give it the madder-bleach, since no white ground has to be preserved. Certain modifications, too, are introduced; it is found, for example, that singeing, and the application of bleaching-powder which causes the formation of oxycellulose, interfere with the production of the most brilliant colour. The apparatus employed being similar to that already described, it is only necessary to give the following summary of the operations usually carried out :—

1. Wash.
2. Boil in water for two hours and wash.
3. Ley-boils 1st, 90 litres (20 gals.) caustic soda. 70° Tw. (Sp. Gr. 1·35), boil ten hours and wash.
    2nd, 70 litres (15 gals.), ditto, ditto.
4. Sour : Sulphuric acid, 2° Tw. (Sp. Gr. 1·01), steep two hours.
5. Wash well and dry.

The above quantities of materials are intended for 2,000 kg. = 2 tons cloth, with low-pressure kier.

In market-bleach the essential difference consists in the absence of the boiling with resin-soap, and the introduction of tinting the cloth with some blue colouring matter previous to drying. With many bleachers, the operation of chemicking comes between the two ley-boils, and not after them, as is usually the case.

Bleaching by Electrolysis.—The production of bleaching liquor by electrolysis is yet somewhat rare in this country, but has already been largely adopted on the Continent. It offers distinct advantages where electric current is cheap. A common salt solution can be split up by the electric current, and chlorine gas given off at the anode of a cell whilst metallic sodium is deposited at the cathode. The sodium is reacted upon by the solution and dissolved, so that sodium hydrate in solution is obtained and hydrogen gas liberated. At the anode the evolved chloride combines with this sodium hydrate, providing the liquor is kept in circulation and the poles are not too far apart, and in this manner is produced liquid sodium hypochloride.

In the Oettel electrolyser a long cell is subdivided into numerous cells by carbon, this latter material being used for both anode and cathode. The cells have a common opening at the bottom, so that when the electric current is turned on the liquid effervesces, and flows over the top into pipes which lead it into the lower and open part of the large cell. In this way the necessary circulation is kept up without the aid of pumps. In the Mather and Platt electrolyser platinum alloy is used for the electrodes, and in this case a pump is necessary to circulate the liquid. A very useful feature of the latter apparatus is the detachable mounting of alternate electrodes, which makes cleaning both easy and possible during working.

CHAPTER VI

LINEN BLEACHING.

THE bleaching of linen is more or less similar to that of cotton, although it is decidedly more tedious, owing to the larger proportion of natural impurities present in the flax fibre, and the greater difficulty of removing or discolorising them. These impurities consist principally of the brown insoluble pectic acid, which remains on the fibre after the retting process, to the extent of 25-30 %. Linen is bleached in the form of yarn, thread, and cloth.

Bleaching Linen Yarn and Thread.—Linen yarn is frequently only partially bleached, and one distinguishes yarns which are " half white " (cream), " three-quarters white," and " full white."

The following is an outline of the general method of bleaching linen yarn as at present adopted in Ireland. The percentages relate to the weight of yarn under treatment :—

1. Boil: 10% soda-ash, boil 3-4 hours; wash and squeeze.
2. Reel: bleaching powder solution, $\frac{1}{2}°$ Tw.; reel 1 hour; wash.
3. Sour: sulphuric acid, 1° Tw., steep 1 hour; wash.
4. Scald: 2-5% soda-ash, boil 1 hour; wash.
5. Reel as No. 2; wash.
6. Sour: as No. 3; wash well and dry.

At this stage the yarn should be " half white."

If it is required " three-quarters white," the drying is omitted, and operations 4, 5, and 6 are repeated with the following slight modifications (*a*) after the "scald" the yarn is " grassed," *i.e.* spread on the grass in a field for about a week; (*b*) instead of reeling the yarn in the solution of bleaching-powder, it is simply steeped in it for 10-12 hours, an operation which is analogous to the chemicking of cotton yarn, and usually called the " dip."

If the yarns should be " full white " the same operations are again repeated once or twice, the duration of grassing being varied according to necessity and the weather. In each succeeding operation the concentration of the solutions employed is diminished.

The operation of boiling takes place in ordinary open or low-pressure kiers, while those of dipping, souring, and washing are best performed in the apparatus illustrated in Fig. 20, page 72. In many establishments, however, the dipping and souring are effected by simply steeping the yarn in stone tanks filled with the necessary liquids, but owing to the absence of all circulation of the latter, this plan cannot be so effective. The washing is frequently done in wash-stocks, or dash-wheels, but this also is not good, because it tends to make the yarn rough.

The mode of applying the bleaching-powder solution in the earlier stages by "reeling," is peculiar to linen yarn bleaching. Its primary object has probably been to ensure regularity of bleach, but since the carbonic acid of the air decomposes the calcium hypochlorite more readily by this means, and liberates hypochlorous acid within the fibre, as it were, the bleaching must be more thorough and greatly accelerated.

The reeling machine consists of a large shallow stone cistern holding the solution of bleaching-powder, and provided with a movable framework supporting a number of reels. On these are suspended the hanks of yarn in such a manner that only their lower ends dip into the liquid. Each single reel can be readily detached if necessary, or, by means of a hydraulic lift, the whole framework with reels and yarn can be raised and withdrawn from the liquid, and at once transferred to another and similar cistern for the purpose of washing, etc. Some bleachers use this machine for scalding.

It is said that better results are obtained if the calcium hypochlorite is replaced by the corresponding magnesium compound. Probably the best agent to use would be sodium hypochlorite, since there would then be no formation on the fibre of any insoluble carbonate, and washing might largely replace the souring, in which case weaker acids even than those mentioned could be used. Since calcium chloride is more soluble than the sulphate, it seems likely that where calcium hypochlorite is used, hydrochloric acid would be better as a souring agent than sulphuric acid.

Bleaching Linen Cloth.—The following is an outline of a modern Irish process for 1,500 kg. = 1·47 tons brown linen (lawns, handkerchiefs, etc.), with low-pressure kiers :—

1. Lime-boil: 125 kg. = 275·57 lb. lime, boil 14 hours; wash 40 minutes in stocks.
2. Sour: hydrochloric acid, $2\frac{1}{2}°$ Tw. (Sp. Gr. 1·0125), steep 2–6 hours; wash 40 minutes in stocks; "turn hank," and wash 30 minutes in stocks.
3. Ley-boils: 1st, 30 kg. = 66·13 lb. caustic soda (solid), 30 kg. = 66·13 lb resin, previously boiled and dissolved together in water; 2,000 litres = 440·19 gals. water; boil 8–10 hours; run off liquor, and add—
   2nd, 15 kg. = 33·06 lb. caustic soda (solid), dissolved; 2,000 litres = 440·19 gals. water, boil 6–7 hours; wash 40 minutes in stocks.
4. Expose in field 2–7 days, according to the weather.
5. Chemick: chloride of lime solution, $\frac{1}{2}°$ Tw. (Sp. Gr. 0·0025), steep 4–6 hours; wash 40 minutes in stocks.
6. Sour: sulphuric acid, 1° Tw. (Sp. Gr 0·005), steep 2–3 hours; wash 40 minutes in stocks.
7. Scald: 8–13 kg. = 17·63 to 28·65 lb. caustic soda (solid) dissolved, 2,000 litres water, boil 4–5 hours; wash 40 minutes in stocks.
8. Expose in field, 2–4 days.
9. Chemick: chloride of lime solution, $\frac{1}{4}°$ Tw. (Sp. Gr. 0·00125), steep 3–5 hours; wash 40 minutes in stocks.

The goods are examined at this stage; those which are sufficiently white are soured and washed, and those which are not are further treated as follows :—

10. Rub with rubbing boards and a strong solution of soft soap.
11. Expose in field 2–4 days.
12. Chemick: chloride of lime solution, $\frac{1}{8}°$ Tw. (Sp. Gr. 0·0006), steep 2–4 hours; wash 40 minutes in stocks.
13. Sour: sulphuric acid, 1° Tw. (Sp. Gr. 0·005), steep 2–3 hours; wash 40 minutes in stocks.

When the cloth (cream linen) is made of yarn already partially bleached, a less severe process is required, *e.g.* less lime is used in the lime-boil; only one ley-boil is given, and that with resin-soap instead of caustic soda; weaker chloride of lime solutions are used; the scald is effected with soda-ash solution; and operations 8 and 9 are omitted.

What is known as "brown holland" is a plain linen cloth which has had little or no bleaching, but only a short boiling in water, or in weak soda-ash solution, followed by a weak souring. It possesses, therefore, more or less the natural colour of the retted flax fibre.

The washing is usually effected in the wash-stocks, but sometimes, and with advantage, too, in so-called slack-washing machines, the washing trough being divided by wooden spars into several compartments, each capable of holding several yards of slack cloth between each nip.

The "rubbing" referred to is a characteristic feature in linen cloth bleaching, and has for its object the removal of small particles of brown matter called "sprits." It is effected by a special machine, which consists essentially of a pair of heavy corrugated boards resting on each other; the upper one is moved lengthwise to and fro, while the pieces are led laterally between them.

The exposing of the goods in a field to the influences of air, moisture, and light, or "grassing," as it is technically termed, is still very generally adopted in order to avoid steeping too frequently in solutions of bleaching-powder, and thus to preserve as much as possible the strength of the fibre.

"Turn-hanking" consists in loosening the entangled pieces and refolding them, so that every part may be exposed to the action of the hammers of the wash-stocks; the operation is introduced as often as required at various stages of the bleaching process, but especially after washing in the stocks.

Chemistry of Linen Bleaching.—During the several boilings with lime, soda-ash, caustic soda, or resin-soap, the insoluble brown-coloured pectic acid of the retted fibre is decomposed, and changed into metapectic acid, which combines with the alkali to form a soluble compound. According to the origin of the flax, the loss which it thus sustains may vary from 15-36 %.

By the application of bleaching-powder alone, the brown pectic matters are bleached only with great difficulty, and even then only by using "chloride of lime" solutions of such concentration that the fibre itself is apt to be attacked. After a number of successive boilings with alkali, however, the goods retain merely a pale grey colour, which is readily bleached by comparatively weak solutions without injury to the fibre.

The rational mode of bleaching linen would seem to be, therefore, to defer the application of the bleaching-powder until the pectic matters have been almost or entirely removed by lime and alkaline boilings, although, in practice, this plan is never strictly adhered to. A single ley-boil scarcely removes more than 10 % of the pectic matters, and since their presence in such large proportion prevents the solutions of bleaching-powder from decolorising the whole of the grey matter at one operation, the usual plan is to alternate the alkaline boilings with dilute chloride of lime treatments; the more so because it is considered that a slight oxidation of the pectic matters facilitates their removal by the alkaline boilings, and that the latter predispose them to oxidation.

Well-bleached linen ought not to be discoloured when steeped in a dilute solution of ammonia; if by this treatment the linen acquires a yellow tint, it is a sign that the pectic matters have not been entirely removed.

The usual period required for bleaching brown linen varies from 3-6 weeks.

CHAPTER VII.

MERCERISING.

The Principles of Mercerisation.—If cotton or linen, in either the raw or manufactured state, is subjected to a cold solution of caustic soda of a density of about 25° Bé., it instantly shrinks, and at the same time becomes stronger and capable of more easily absorbing dyestuffs than in its original state. If, however, the material is treated whilst under tension so as to prevent shrinkage, or if whilst still wet it is stretched back to as near its original dimensions as possible, the fibres swell and become translucent, as when treated without tension, but at the same time there is a drawing action amongst them which ultimately leaves them with a gloss not unlike that of the silk fibre.

Experience has clearly shown that solutions of soda-lye stronger than 25° Bé. do not produce any better effect, and that also it is useless prolonging the immersion, for the full possible amount of action takes place in about a minute. With weaker solutions than the above a corresponding weaker lustre is obtained, while little effect is possible with lyes weaker than about 20° Bé. Heating the lye has no better effect unless the solution is made correspondingly strong, about 50° C. being the best working temperature. The mercerising action is discernible on all classes of cotton and linen, but it is not found advantageous to work any but the long-fibred cottons, as with the short-fibred classes the result is insignificant.

Tensioning Devices.—It is extremely difficult to properly mercerise raw material, as there is no reliable method of gripping the fibres and keeping them under tension. The best way is to lay the cotton or linen between two sheets of fine mesh wire fabrics, and whilst held between these to treat with soda lye and wash off.

Yarn is the most satisfactory state in which to mercerise

vegetable fibres, especially when it is in hank form. The hanks are placed over two rollers, and these can be forced apart by either levers or by screws. Frames carrying these rollers can be filled with hanks and dropped into the different vats: first, one containing soda lye; second one containing clean water for washing-off purposes; third, a souring tank containing a weak solution of sulphuric acid to neutralise any remaining caustic soda; and fourth, another wash to clear away the acid.

Not only are there various frames of this description on the market, but there are several good machines which automatically perform the operations. These are a great advantage, for the process can be carried out very rapidly, whilst special gloves need not be worn, although it is necessary when the operator has to handle the yarn before the lye is washed out.

Yarn can also be mercerised whilst in warp form. The warp is passed through what is practically a warp dyeing machine, but the warp is kept tight between the sets of rollers, so as to prevent shrinkage.

The mercerisation of piece goods is usually carried out by treating the woods with the lye and then stretching them back to width on the tentering machine. In other cases the stretching is done in the crabbing machine, but it will be readily seen that the former will only stretch the weft threads properly, whilst the latter only influences the warp. There are also special tentering machines made in which the cloth can be taken through the several baths and prevented from shrinking during the process, instead of the after-stretching above-mentioned.

Fancy Effects Obtained by Mercerisation.—Fancy effects can be obtained by mercerisation by using either the lustring properties of the process or the shrinkage which is present. By the former, invisible effects may be obtained by printing soda lye on a fabric under tension, when the figures come out in the same shade as the rest of the piece, but brighter. This applies only to white goods, for the parts so printed would show a darker shade in a dyed piece.

Effects may be also obtained of a crêpon nature by printing portions of a fabric and allowing it to shrink, whilst the untreated parts retain their usual dimensions.

MERCERISING.

Lustring Without Tension.—Attempts have been made to obtain the lustre which follows treatment in soda lye under tension without the accompanying tension. So far the methods attempted have consisted of adding something to the lye, but although there has been a small measure of success it is usually found that the lustre suffers.

The glycerine process is so far the most successful, one part of glycerine mixed with two parts of soda lye 50° Bé., practically preventing shrinking and giving a lustre which almost comes up to properly treated goods.

The other methods already tried consist of adding alcohol, gelatine, waste silk, dissolved wool, glucose, sodium silicate, hydrocarbons, etc., to the caustic soda.

CHAPTER VIII.

WOOL SCOURING AND BLEACHING.

Object of Scouring Wool.—The "scouring" of wool has for its object the complete removal of all those natural and artificial impurities (yolk, dirt, oil, etc.) which would otherwise act injuriously during the processes of weaving or dyeing.

The paramount importance of having woollen material thoroughly well cleansed or scoured previous to dyeing, cannot be too much insisted upon. If the operation be omitted, or be improperly or incompletely performed, each fibre remains varnished, as it were, with a thin layer of fatty matter, which resists to a greater or less degree the absorption and fixing of mordant or colouring matter. The final result is that the material is badly or unevenly dyed, and appears "flecked," "stripey," etc., or since the colouring matter can only be superficially deposited, it is readily removed by the subsequent operations of washing, milling, etc.

It has already been mentioned, when considering the wool fibre, that raw-wool is impregnated or encrusted with yolk, consisting essentially of certain fatty matters, free fatty acids, etc. Woollen yarn and cloth always contain oil (olive oil, oleïc acid, etc.) to the extent of about 10-15 %, which has been sprinkled on the wool for the purpose of facilitating the spinning process.

It would be a great advantage, however, if manufacturers would more frequently use neutralised sulphated castor-oil soap solutions instead of oil, for this purpose. Cloth which required milling after dyeing (*e.g.* mixtures, tweeds, etc.), or previous to dyeing, would then need no scouring with alkali, but merely a simple washing in soft water. This plan has been adopted with success in Belgium.

The method of removing yolk, oil, etc., from wool,

which first suggests itself, is that of saponifying or emulsifying them by means of weak alkaline solutions, and such, indeed, are the agents most largely employed for the purpose of wool scouring.

The choice of the most suitable alkaline scouring agent and the best temperature to be employed depend upon the quality and origin of the wool.

In modern times other agents of a volatile nature, and capable of dissolving fatty substances, have been proposed and to some extent employed—*e.g.* fusel oil, ether, light petroleum oil, carbon disulphide, etc.—and it is by no means improbable that some member of this class will form the wool-cleansing agent of the future. It must be remembered, however, that these volatile liquids are essentially solvents for fatty matters, and not for alkaline oleates. A washing in water would therefore always have to take place in addition.

Scouring Agents.—Lant or stale urine is the detergent which has been employed for wool-scouring from the earliest times, and it is effective because of the ammonium carbonate it contains. Though still used, this agent has been largely supplanted by others, such as ammonia, sodium carbonate, soap, etc.

Stale urine, used in the proportion of 1 measure to about 5 of water, gives excellent results in most cases. It leaves the wool clean and with its normal physical properties of softness, elasticity, etc., unimpaired, its disagreeable odour being its main defect. Ammonium carbonate is the most rational substitute for "lant," and represents, perhaps, one of the best alkaline detergents, especially if used along with soap; but it is still too expensive for general use. As a mild scouring agent for the better classes of wool, potash or soda soap is largely employed. Satisfactory results are obtained with these if the soap is of good quality, and free from excess of caustic or carbonated alkali. Other things being equal, the most soluble soaps are to be preferred, such as potash soaps, or soda soaps made from oleïc acid. An average soap solution may contain 30-50 g. of soap per litre of water (4·8-8 oz. per gal.). An addition of ammonia to the soap solution is frequently made for the purpose of increasing its detergent properties. An important consideration to be remembered when using soap is, that the water should be as free as possible from lime and magnesia to avoid the formation of lime and magnesia soaps.

The scouring agent most largely employed, either alone or mixed with soap, is sodium carbonate, of which excellent qualities (Solvay soda, "Crystal carbonate," etc.) are now to be had. When used with judgment and care, it leaves the wool but little affected, and, being inexpensive, it is well adapted for wools of medium and low quality. The injurious effects of using a calcareous or magnesian water are less marked with this agent than with soap, since the calcium and magnesium carbonates precipitated on the wool are of powdery nature and more readily removed. The presence of caustic soda must always be rigidly avoided. The concentration of the soda solution should be such that it stands at 1°-2° Tw. (Sp. Gr. 1·005-1·01), or, say, 10-20 g. = 0·35-0·70 oz. $Na_2CO_3 10H_2O$ per litre. It may be stated, as a general rule, that the best results are obtained (*i.e.* as regards feel and lustre of wool) by using the solutions of alkaline detergent as dilute, and at as low a temperature as is consistent with the complete removal of all impurities. The temperature may vary from 40°-55° C.

Other substances have from time to time been recommended as useful additions to the scouring bath, *e.g.* common salt, ammonium chloride, decoction of soap bark (*Quillaya saponaria*), oleïn, resin soap, pigs' manure (Seek), etc. Some of these may act beneficially in certain cases, but as a rule the previously mentioned scouring agents are more generally useful. Ammonium chloride may de-alkalise soap, or if used along with sodium carbonate may be beneficial because of the ammonium carbonate produced. Oleïn and resin-soap probably assist emulsification, while seek, which is frequently employed, is effective because of its alkalinity.

Patent or secret scouring substitutes should be strictly avoided; they are either worthless, or, if useful, they can be prepared by the scourer himself more economically. Secrets, both in scouring and in dyeing, belong rather to the past than to the present age.

Scouring Loose-Wool :—

a. Scouring with Alkaline Solutions.—In its most complete form the scouring (Fr. *lavage*) of raw or greasy wool should consist of at least three operations :—

1st. Steeping or washing with water (Fr. *désuintage*).

2nd. Cleansing or scouring proper with weak alkaline solutions (Fr. *dégraissage*).

3rd. Rinsing or final washing with water (Fr. *rinçage*).

In England the first operation is omitted, sometimes because the wool has been already more or less completely washed by the wool grower, in order to lessen the expense of transport, but more frequently because the opinion prevails that the operation is unnecessary and only entails additional expense. It is contended, and indeed with some degree of truth, that the soluble portion of the yolk, owing to its alkaline nature, renders material assistance in the scouring bath.

On the other hand, this system has certain defects. The scouring bath becomes more readily soiled and rapidly varies in composition, so that, unless it be frequently renewed, it soon scours the wool insufficiently, or may even stain it injuriously. Under ordinary circumstances, and scouring with sodium carbonate or with soda soaps, the yolk is entirely lost.

When the wool contains only a small percentage of yolk, the steeping process for the recovery of yolk-ash may well be omitted, but its adoption in the case of wools rich in yolk (as those from Buenos Ayres, etc.) is certainly to be recommended as distinctly advantageous. In Belgium and France it is largely carried on.

Steeping.—To be effective the process must be systematic, the water should be heated to about 45° C., and the wool if possible agitated.

In practice, the last point is not attempted on economical grounds, and the wool is simply steeped in, or rather systematically washed with, tepid water in the following manner :—

Four or five large iron tanks capable of being heated by steam are filled with the raw wool. The first one is then filled with tepid water (45° C.), and the wool is allowed to steep for several hours until, indeed, no further quantity of yolk is dissolved; by means of a siphon, steam-injector, pump, or other means, the liquid is removed to the second tank, where it is also allowed to remain (at 45° C.) until it again ceases to dissolve yolk; it is then in like manner removed to tanks 3 and 4, until finally it becomes perfectly saturated with yolk, and is ready for being evaporated to dryness, calcined, etc., to obtain the yolk-ash. Simultaneously with what has just been described, *fresh* water (45° C.) is made to circulate in a similar manner through the several tanks, until at length the wool in tank No. 1 is entirely deprived of the soluble constituents of the yolk; the wool being now ready for the scouring. This tank is emptied and refilled with raw wool. So soon as this has taken place, the steeping water is made to circulate through the tanks in the order of Nos. 2, 3, 4, 1, so that the saturated yolk solution is now drawn off from the newly-filled No. 1 tank, and the wool contained in No. 2 tank is exhausted, which is then duly emptied and refilled.

By the above explained method of steeping the wool in each tank becomes, in turn, gradually deprived of yolk, and supposing the tanks to be arranged in a circle, the points at which the fresh water is introduced and the saturated yolk solution withdrawn, are always two contiguous tanks, and move gradually round the circle. The main principle of the process is, that the raw greasy wool is always first washed with water already pretty well saturated with yolk, the partly extracted wool is brought into contact with weaker yolk solutions, while the most thoroughly exhausted wool is washed with fresh water.

Figs. 26 and 27 represent an arrangement of washing tanks designed by H. Fischer with a view to economise space and labour.

A, B, C, D, represent four iron tanks suspended between two large wheels F, having the common axis E. One of the wheels is provided with teeth, which work against the small cog-wheel of a windlass, the arrangement being such that the whole apparatus can be readily turned at G by a single workman. Each tank can thus be raised in turn to a high level for the purpose of draining the wool it contains and running the liquid into the next tank.

The following numbers giving the amount of dry yolk in yolk solutions of different degrees of concentration are calculated from data published by Havrez:—

Spec. Gravity of yolk solution.	Degrees Twaddell.	Amount of dry yolk contained in one litre.
1·377	75·4	769 grams.
1·215	43	376
1·102	20·4	175
1·048	9·6	72
1·025	5	37
1·012	2·4	17
1·006	1·2	8

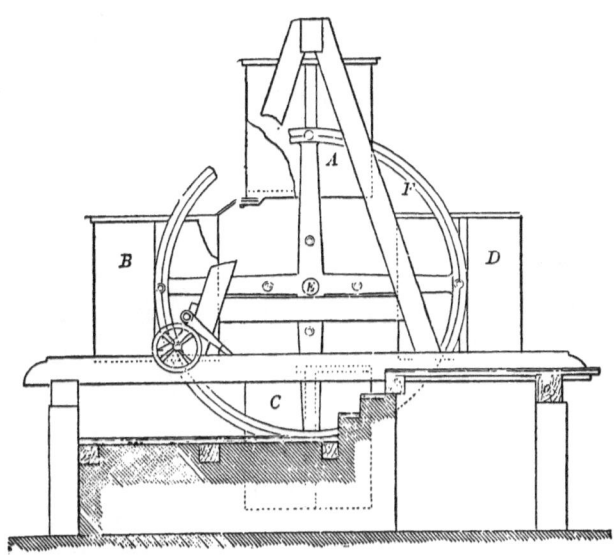

Fig. 26.—Wool-steeping Tank—Elevation.

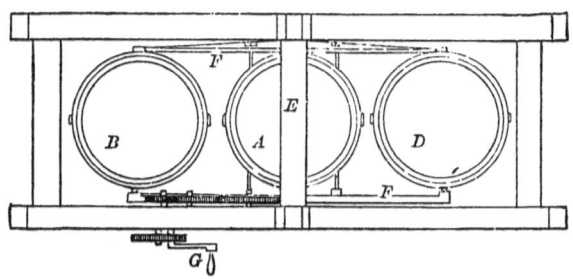

Fig. 27.—Wool-steeping Tank—Plan.

According to M. Chandelon, 1,000 kg. = 2,204 lb. of raw wool may furnish 313 litres = 68·88 gal. of yolk solution of Sp. Gr. 1·25 (50° Tw.), having a value of 15s. 6d., while the cost of extraction does not exceed 2s. 6d.

Fig. 28 represents a furnace for the manufacture of carbonate of potash from yolk, devised by H. Fischer.

The concentrated yolk solution is first put into the tank a and warmed by the waste heat of the furnace. Sometimes tank a is situated at a high level, and the concentrated yolk solution is led into the chimney, where it trickles down in a ziz-zag course across overlapping steps at once into the chamber b. The flow can be so regulated that the level of the liquid in b is constant. In the regular course of work the liquid is run from the

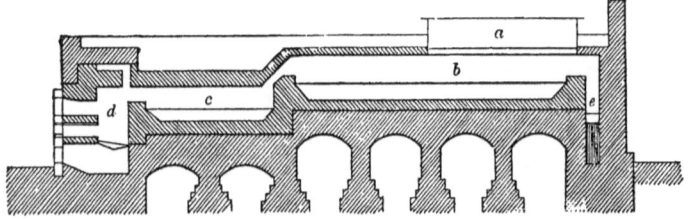

Fig. 28.—Section of Furnace for Making Yolk-ash.

tank a into the evaporating chamber b, and when partly evaporated here, it is led by a connecting pipe into the calcining chamber c, where it is evaporated down to a thick syrupy consistency. Very soon, owing to the organic matter contained in it, it ignites, and the fire at d can be much moderated. The calcined mass in c is worked about with a rake until it acquires a dirty grey colour, when it is withdrawn and allowed to cool. In the meantime, the liquid in b has been evaporating, and fresh liquid has been warmed in a. When properly managed the process is almost continuous. Care must be taken in admitting new liquid from b into the red-hot chamber c, since it froths up very considerably. The pipe connecting b and c is so arranged that it can be readily removed and cleaned. Comparatively little coal is required, since it is largely economised through the burning of the organic matter of the yolk itself. It has been found in practice that to evaporate 12 kg. = 26·44 lb. of water in such a furnace 1 kg. = 2·2 of coal is required. Yolk ash is recovered in Roubaix, Antwerp, Verviers, Louvain, Brugge, Hanover, Döhren, and Bremen.

Scouring and Washing.—In small establishments the wool is thrown, without previous steeping, into a large tank filled with the scouring liquid, and worked about by hand for a short time with poles. It is then lifted out with a fork, drained on a wooden screen, and well washed several times in a cistern having a perforated false bottom. When soap is used, the excess of liquid is removed by a pair of squeezing rollers before washing. Obviously, by this method the wool is only incompletely or irregularly scoured, hence in large and well-equipped establishments, after "steeping," it is passed through the wool-scouring machine.

The great object of the wool-scourer should be to maintain the composition of the second bath as constant as possible, or at least not to allow it to become too much soiled. This is effected partly by the squeezing rollers between the first and second machines, and partly by emptying the first bath the moment it gets charged with fatty matter, transferring to it the slightly-soiled liquor of the second bath, and refilling the latter with fresh scouring solution. No doubt a still more regular and complete scour would be obtained by having a range of four troughs instead of three. If the amount of fatty matter remaining in the wool after scouring exceeds 1%, the operation must be regarded as having been inefficiently performed.

Figs. 29 and 30 show, respectively, elevation and cross section of the McNaught wool-scouring machine. It consists of a long tank A containing the scouring liquor, part of the tank being partitioned off at the top and provided with a perforated bottom B. The wool is fed into the machine by the lattice C, and, after passing through rolls, is delivered into the upper compartment of the tank A, along which it is passed by forks D. These forks are swung from the framework shown, and by means of cams are operated so as to dip into the liquor, have a movement in the direction in which the wool should travel, lift from the liquor, and have a backward or return movement in mid-air. They are thus continually propelling the wool

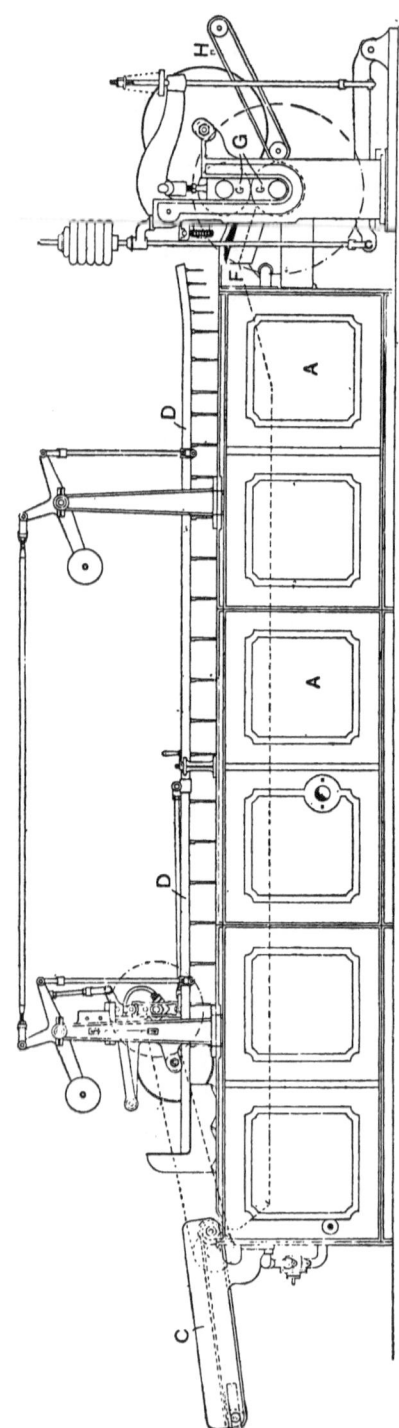

Fig. 29.—McNaught's Wool-scouring Machine—Elevation.

WOOL SCOURING AND BLEACHING.

in one direction, whilst the return current of liquor can flow along the upper part of the tank A, outside the compartment containing the wool. The dirt which is loosened from the wool falls through the perforated bottom plate B and slides down the inclined bottom of the tank A, from whence it is cleaned at intervals.

On reaching the right end of the drawing, the wool is forked up an incline F, and then falls down another incline into the nip of the squeeze rolls G, which pass it to

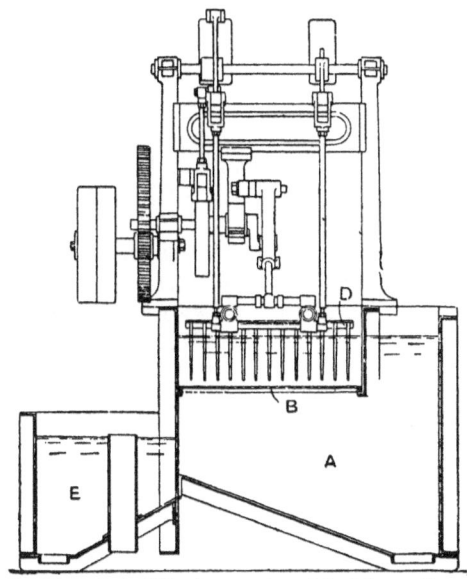

Fig. 30.—McNaught's Wool-scouring Machine—Cross Section.

the lattice H. It then goes forward to a similar machine, or may be passed forward to be dried. The liquor ejected by the squeeze rolls G runs through perforations in the incline and down to the small tank E (Fig. 30), from whence it is pumped back again into the main tank A.

A really effective scouring arrangement consists of at least three such machines, placed in line, so that the wool may be passed automatically from one to the other. The operation of scouring and washing thus becomes continuous, regular, and complete.

The first machine into which the wool is introduced contains more or less soiled scouring liquor, which has already been used in the second trough; the latter contains fresh scouring liquor, and the third a continual flow of clean, cold, or preferably tepid, water.

Magma Process.—The waste scouring liquor ought to be collected in stone-lined pits, and there neutralised or acidified with sulphuric acid. The magma of fatty matter which rises to the surface is collected, drained in filter bags, and sold to oil dealers. If soap has been the scouring agent employed, the fatty acids thus recovered are all the more valuable, but in any case the spent scouring liquors should never be allowed to pollute the neighbouring stream.

b. Scouring with Volatile Liquids.—This is more advantageous than the alkaline method, because it deprives the wool more completely of its wool-fat, and the injurious effect of the alkalis is entirely removed. On the other hand, the method is more costly, and by the use of *some* extracting liquids the wool may certainly be modified.

Some consider that since oil must be added to the wool before spinning, it is not necessary to remove the whole of the natural fatty matters from raw wool. Whether this view be correct or not as far as spinning is concerned, it is certainly not to be entertained if the loose wool has to be dyed.

Up to the present time, wool scouring with volatile liquids has not met with general acceptance, partly because of the attendant danger if not employed with great care and with suitable appliances. The difficulties, however, are not insuperable, and have, indeed, been more or less overcome by Da Heyl, Van Haecht, and others.

One method is that of T. J. Mullings. The wool is placed in an enclosed centrifugal machine, and submitted to the action of disulphide of carbon. When this liquid is saturated with yolk, the machine is set in motion to remove the bulk of it, the remaining portion being expelled by admission of water. The wool is afterwards washed with water in the usual washing machines. The novel feature in the process is the expulsion of the carbon disulphide by displacing it with water, by which means the wool does not acquire the yellow tint it invariably assumes when heat is employed for this purpose. The mixture of carbon disulphide and water is collected in a tank; after settling, the former is drawn off from below and recovered for subsequent use by distillation. Experiments are said to have shown that wool cleansed in this way is stronger, and

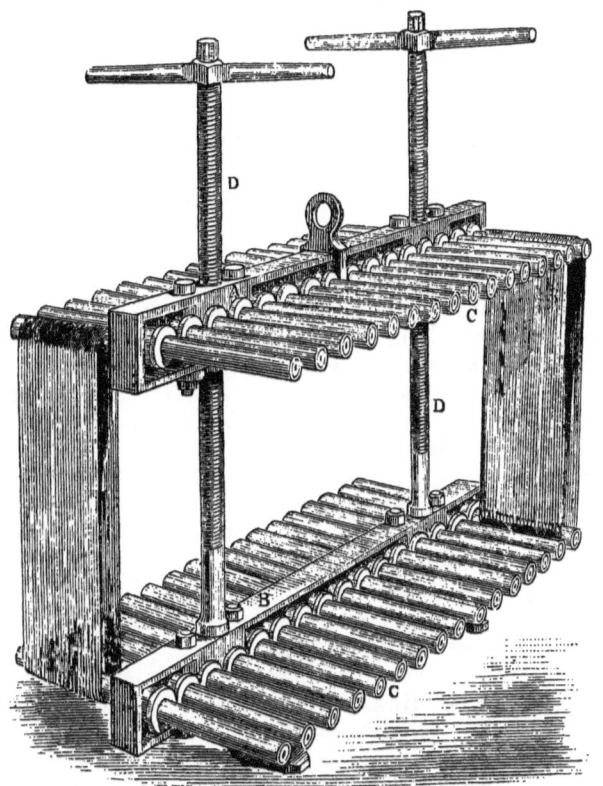

Fig. 31.—Yarn-stretching Machine.

will spin finer yarn, and with less waste, than if scoured by the ordinary method with soap, and this is done at one-eighth of the usual cost.

Yarn Scouring.—The scouring of woollen yarn is more readily effected than that of raw wool if the oil with which it has been impregnated by the spinner has been of good quality (*e.g.* olive oil). It has always been considered that the difficulty is increased if cheap oils have been used which contained an appreciable amount of mineral oil, since this, being a hydro-carbon and not a glyceride, is unsaponifiable. Experiments by C. Roth seem to show, however, that this view is erroneous.

a. Stretching of Yarn.—Those yarns which are hard twisted require this preliminary process in order to remove their curly appearance and to prevent them from shrinking during the subsequent scouring operation. For this purpose the " yarn-stretching " machine (Fig. 31) is employed.

It consists of two vertical iron screws D, connecting two horizontal bars, A, B, one above the other, each fitted with a series of metal pegs or arms C, on which the hanks are suspended. The lower bar B is fixed, while the upper one A is capable of being moved up or down by turning the vertical iron screws, and fixed at any point. After filling the arms C with yarn, as indicated in the figure, the bar A is screwed up until the yarn is suitably stretched. In this condition the whole apparatus is immersed in a bath of boiling water, and after a few minutes removed. Those portions of the hanks immediately in contact with the pegs are still unaffected and look curly. Hence after relaxing the tension of the hanks, their positions on the pegs are changed, they are again screwed up, and the immersion in boiling water is repeated. When taken out and allowed to cool, the yarn is taken off ready for scouring.

b. Scouring Yarn.—Yarn scouring is generally done by hand in an ordinary rectangular wooden tank, the liquor being heated by means of a perforated copper steam-pipe. The hanks are suspended on smooth wooden rods placed across the tank, on each side of which stands a workman. One by one the rods full of yarn are taken up, once or twice swayed to and fro, and then each hank is carefully lifted up and turned, so that the exposed portion resting on the rod may become immersed. This is frequently facilitated by means of a second and thinner rod, which is inserted in the loop of the hanks, immediately beneath the suspending rod, so that the whole rod full of hanks may be turned at once, and without scalding the hands. The whole operation is systematically repeated during 15-20 minutes, after which the yarn is transferred to a second tank containing cleaner scouring liquid. Here the process of turning is repeated, after

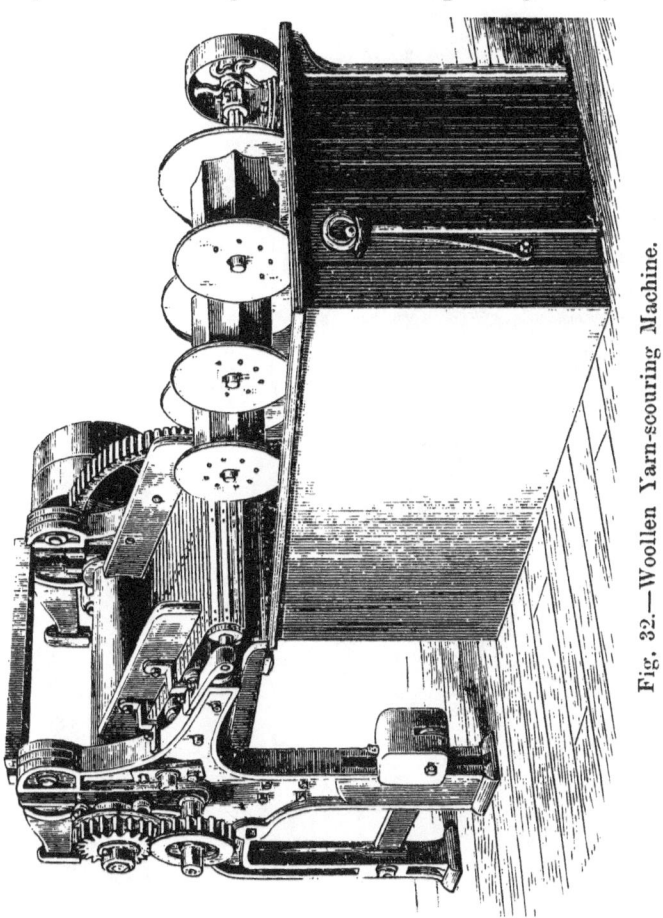

Fig. 32.—Woollen Yarn-scouring Machine.

which the yarn is washed, either by the same method or by placing the rods full of hanks on a pair of horizontal bars, situated beneath a perforated wooden tray, **on which** water is flowing; the yarn thus receives an efficient shower bath.

Scouring partly by hand and partly by machine is effected by the apparatus represented in Fig. 32. The

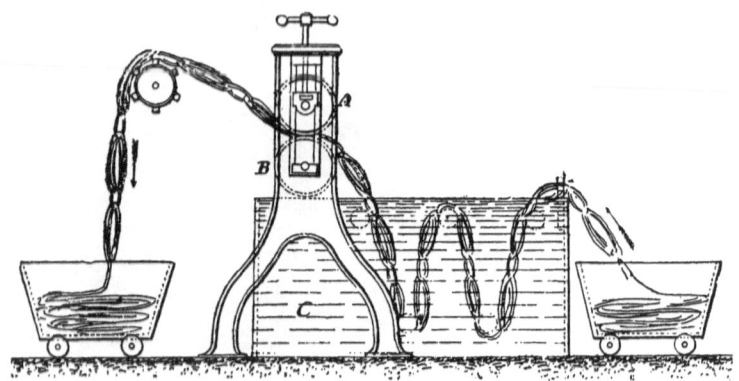

Fig. 33.—Continuous Woollen Yarn-scouring Machine.

Fig. 34.—Woollen Cloth-scouring Machine.

yarn is suspended on reels projecting from one side of the scouring box, and caused by steam-power to revolve in the scouring solution alternately in each direction for a short time. The hanks are then taken off the reels, placed on a moving endless band, and thus led through a pair of squeezing rollers, to be washed with water in a similar machine. In some machines the reels are omitted, and the hanks worked in the liquid by hand.

A perfectly continuous method of scouring by machinery alone is that in which the loose hanks of yarn are placed on a feeding apron, and borne along between two broad endless bands, through a succession of scouring baths fitted with a series of squeezing rollers.

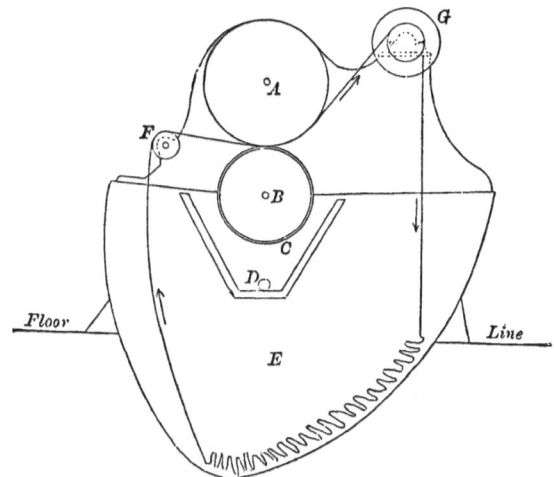

Fig. 35.—Section of Machine shown in Fig. 34.

Another continuous method is that carried out by means of the machine represented in Fig. 33.

The hanks are linked together by means of a small knotted and twisted loop of cord. The chain of yarn thus formed is then passed continuously through a series of three machines, similar to the one represented, c. The squeezing rollers, A and B, are thickly covered with some soft, durable material, such as silk noils.

Cloth Scouring.—Woollen cloth is either scoured in the "rope" form or in the open width, the latter being the preferable mode, since the operation cannot but be thus more evenly performed, and does not tend to produce creases in the material.

Figs. 34 and 35 represent (in perspective and in section) the common rope-scouring machine, usually termed a "dolly." It consists essentially of a pair of heavy

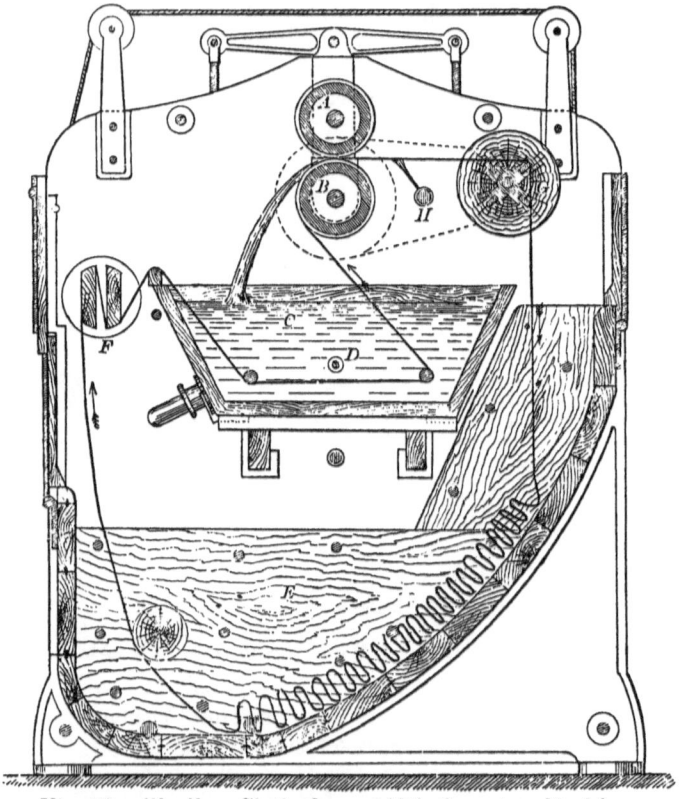

Fig. 36.—Woollen Cloth Open-width Scouring Machine.

wooden squeezing rollers (A B), placed over a box or trough C, containing the scouring solution. The pieces are stitched end to end, to form an endless band, and this is made to pass continuously for twenty minutes or so between the rollers, the lower one of which, since it is partially immersed, carries the solution to the cloth.

D is a steam pipe for heating the scouring solution, E an empty wooden box, F and G are guiding rollers. The operation is repeated with fresh solution in another similar machine, and the pieces are afterwards washed with clean water.

The machine employed for scouring cloth in the open width is shown in Fig. 36. It is much broader than the one just described, in order to suit the width of the cloth, and certain straining bars F are required for the purpose of keeping the cloth opened out and free from creases previous to its passage between the squeezing rollers A, B, which are preferably of iron, in order to give a more equal pressure across the whole width of the cloth. Immediately below these rollers is the trough O containing the scouring solution. The perforated water-pipe H is used when the pieces are washed in this machine after scouring.

Scouring Union Material.—The scouring of thin materials with cotton warp and woollen weft presents certain difficulties. The different hydroscopic, elastic, and other physical properties of cotton and wool, cause such materials, if simply scoured in the ordinary way, to contract or shrink irregularly over the whole surface of the fabric, so that they assume, when dried, a rough, shrivelled appearance which renders them quite unsaleable. Special appliances and methods are in consequence required. The scouring of thin union goods comprises the operations of Crabbing, Steaming, and Scouring.

(a) *Crabbing or Fixing.*—The object of this and the following operation is to prevent the material from acquiring the "cockled," "curled," or shrivelled appearance above alluded to. It also imparts a permanent and indestructible lustre and finish of a peculiar quality, which is not removed or affected by any subsequent operation.

Fig. 37 shows the arrangement of a treble crabbing machine.

The cloth A wrapped on the roller or beam B is passed in the open width, and in a state of tension, below the roller D, and through the boiling water contained in the vessel C, then immediately between the pair of heavy iron rollers B and E, under great pressure. It is at once

114 TEXTILE FABRICS.

Fig. 37.—Treble Crabbing Machine.

tightly wrapped or beamed on the lower roller B, while still revolving in the hot water. The process is repeated with boiling water in the second trough, and again with cold water in the third trough. The tension of the cloth and the pressure of the rollers are varied according to the quality of the goods, and the particular feel, lustre, and finish ultimately required. With goods that must have subsequently a soft feel or "handle," such as Cashmeres, Coburgs, etc., pressure is not employed, the pieces being simply beamed tightly on the bottom roller.

(b) *Steaming.*—The pieces are unwrapped from the last crabbing roller (that is, from the cold water), and tightly wrapped on the perforated revolving iron cylinder G. Steam is admitted through the axis of this cylinder for the space of about ten minutes, or until it passes freely through the cloth. In order to submit every portion of the piece to an equal action of the steam, the process is repeated with the cloth tightly beamed on a second and similarly perforated roller G', so that those portions of the cloth which were on the outside are now in the interior. These perforated steam-cylinders are frequently quite separated from the crabbing machine, and then usually rest on a steam nozzle in a vertical or horizontal position.

(c) *Scouring.*—The cloth, being now "set," as it is technically termed, is scoured for half an hour or more, with soap solution at 40°-50° C. in the "Dolly" or "open width" machine, above described.

The sequence of operations as here given, although frequently employed, is not altogether rational.

The best results are obtained by crabbing and scouring simultaneously, and then steaming. To accomplish this, the boiling water in the crabbing troughs is merely replaced by a solution of soap, sodium carbonate, or other scouring agent. It is found that to steam the cloth in its oily state exercises some injurious action, and renders it liable to contract dark stains during mordanting, especially if stannous mordants are employed. It is possible that the oil is more or less decomposed, and a portion becomes fixed on the fibre.

Bleaching Wool.—Wool is generally bleached either

in the form of yarn or cloth, but only when it is intended to remain white, or if it has to be dyed in very light delicate colours. The bleaching agent universally adopted is sulphur dioxide. According to the state in which it is applied, either in the form of gas or dissolved in water, one may distinguish between "gas bleaching" and "liquid bleaching," and of these the former is more generally employed. In recent years peroxide of hydrogen has come to the front, and is gradually being adopted for this purpose.

Gas Bleaching, Stoving, or Sulphuring.—Yarn is first

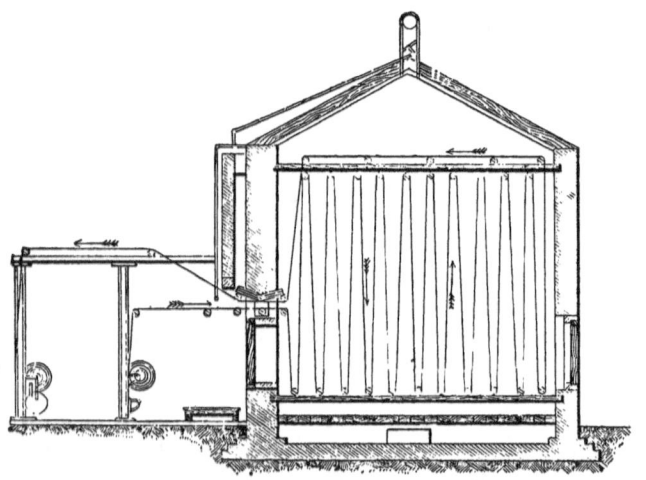

Fig. 38.—Sulphur Stove for Woollen Cloth Bleaching.

scoured and well washed, then suspended on poles and placed in the sulphur stove—a spacious brick chamber which can be charged with sulphur dioxide. The necessary amount of sulphur (in the proportion of 6-8 % of the wool to be bleached) is placed in an iron pot in one corner of the chamber, and ignited by inserting a hot iron; the chamber is then closed, and the moist yarn is left exposed to the action of the gas for 6-8 hours, or even overnight. Afterwards the chamber is thoroughly ventilated; the yarn is removed and well washed in water.

Heavy woollen cloth, such as blanketing, is treated

WOOL SCOURING AND BLEACHING.

in exactly the same manner as yarn, but with thin material—*e.g.* merino, etc.—the operation is preferably made continuous by adopting the arrangement of stove shown in Fig. 38. It is provided internally with a wooden frame, having rollers above and below. The roof should be lined with lead, and heated with steam pipes, in order to prevent condensation. The stove is charged with sulphur dioxide as already described, or, preferably, the sulphur is burnt in a separate furnace, and the gaseous product is led underneath the perforated tile floor of the stove. The cloth is introduced through a narrow slit in the wall; it then passes, as indicated, under and over the rollers, and passes out again by the same opening. The number of times the cloth is passed through the stove varies according to the appearance of the cloth.

In liquid bleaching, the woollen material is worked and steeped for several hours either in a solution of sulphurous acid or in a solution containing sodium bisulphite (5-50 g. per litre), to which an equivalent amount of hydrochloric acid has been added; it is afterwards thoroughly washed. A better method, however, is that in which the wool is treated with sodium bisulphite and by hydrochloric acid in separate baths, whereby the sulphurous acid is generated within the fibre, and, being in the nascent state, acts more powerfully upon the colouring matter of the wool. Goods which have to remain white are tinted with some blue or bluish-violet colouring matter (such as ground indigo, indigo-extract, aniline blue, etc.), either before or after the bleaching operation, in order to counteract the yellow colour of the wool which is so apt to return. The principle here applied is that of the complementary colours, which, when mixed in due proportion, produce white light. Blue is complementary to yellow.

The bleaching action of sulphur dioxide is most probably due to its reducing action upon the natural yellow colouring matter of the wool; another explanation, however, is that it combines with the latter to form a colourless compound. Certain it is that the effect is by no means permanent; frequent washing of bleached wool in alkaline solutions always tends to restore the yellow appearance of the fibre. Either oxidation is thus induced, or the colourless sulphite is decomposed, and the original colouring matter is precipitated on the fibre.

The agent *par excellence* for liquid bleaching is peroxide of hydrogen (H_2O_2). Even yellow-coloured wool is bleached by it to a white, possesssing a brilliancy and purity unattainable by the ordinary methods. The woollen material is steeped for several hours in a dilute and slightly alkaline solution of commercial peroxide of hydrogen and afterwards well washed, first with water acidulated with sulphuric acid, and afterwards with water only.

CHAPTER IX.

SCOURING AND BLEACHING SILK.

Object of Scouring Silk.—The object of scouring silk is to remove from the raw fibre a greater or less proportion of the silk-glue which envelops it, and thus to render it lustrous and soft, and better fitted for the operation of dyeing. According to the amount of silk-glue removed, the product of the scouring operation may be either boiled-off silk, souple silk, or écru, for each of which, indeed, a different treatment is necessary.

Boiled-off Silk is the name given to silk from which practically the whole of the silk-glue has been removed. It exhibits most fully the valued properties of lustre, softness, etc. Two operations are necessary for its production, namely, " degumming " and " boiling-off."

(*a*) *Degumming* (Fr., *dégommage*).—The object of this operation is to soften the silk, and to remove the great bulk of the silk-glue and also the colouring matter. The hanks of raw silk are suspended on smooth wooden rods, and worked by hand in rectangular copper troughs, in a solution of 30-35 % soap, heated to 90°-95° C. When the water is very calcareous, the silk is first rinsed in a weak, tepid solution of sodium carbonate. The best plan is to correct the water previously. Rinsing in dilute tepid hydrochloric acid before degumming is also good, since it removes calcareous and other mineral matters from the silk, and prevents their action in the soap bath. Weighted écru silks cause great inconvenience in the degumming: the soap bath is precipitated, the fibre becomes tarnished and sticky, and the degumming is rendered difficult and incomplete.

During the stripping operation the silk at first swells up and becomes glutinous, but, after a short time, when the silk-glue dissolves off, it becomes fine and silky. It is best, especially when the silk is intended for whites or delicate colours, to work it successively in two or three separate baths, for about 20-25 minutes in each, and to pass fresh lots of silk through in regular order. When the first bath becomes charged with silk-glue it is renewed, and then employed as the last bath. Each soap bath should be utilised to the fullest extent compatible with excellence of result. It is well to bear in mind that too prolonged contact with boiling soap solution is not good, since a little of the colouring matter of the glue is apt to be attracted by the fibre, and the silk loses substance, strength, and purity of white. The waste soapy and glutinous liquid obtained is called "boiled-off" liquor, and serves as a useful addition to the dye-bath when dyeing with the coal-tar colours.

After "degumming," the hanks are rinsed in water, in which a small quantity of soap and sodium carbonate has been dissolved.

(b) *Boiling-off* (Fr., *la cuite*).—For the purpose of removing the last portions of silk-glue, etc., and to give the silk its full measure of softness and lustre, it is now placed in coarse hempen bags, technically called "pockets," about 15 kg. in each, and boiled from half an hour to three hours (according to the quality of the silk) in large, open copper vessels, with a solution of 10–15 % of soap. The silk is then rinsed in tepid water, rendered slightly alkaline by the addition of sodium carbonate in order to prevent the precipitation of lime soap on the silk. It is finally washed well in cold water. The waste soap liquor may be used for "degumming." For some articles, the silk is boiled on wooden rods and not in pockets.

The soap employed, both for "degumming" and for "boiling-off," should be of the best quality. Other things being equal, those soaps are to be preferred which wash off most readily, and leave an agreeable odour. When, however, the silk has subsequently to undergo a number of soaping and other operations—as in weighted blacks—the odour of the soap used in boiling-off is of little consequence. Oleïc acid soap may be recommended in such a case, but for silk destined to be dyed in light colours or to remain white, a good olive-oil soap is preferable.

By scouring with soap in the above manner, Japanese

and Chinese silks lose 18-22 % of their weight and European silks 25-30 %.

Stretching (Fr., *étirage*).—When the silk is well softened during the stripping process, but not really degummed, it may be conveniently stretched to the extent of 2-3 % without injury; indeed, it acquires increased lustre by stretching. The operation may be performed either by the lustring or the stringing machine.

Stoving.—When the silk is intended to remain white, or is to be dyed pale colours, it is bleached by exposing it in a moist condition to the action of sulphurous acid in closed chambers. The operation, which lasts about six hours, may be repeated 2-8 times, according to the nature of the silk. Ten kg. of silk will require about $\frac{1}{2}$ kg. of sulphur. After stoving the silk is well washed till free from sulphurous acid.

Souple silk is silk which has been submitted to certain operations, to render it suitable for dyeing, etc., without causing it to lose more than 4-8 % of its weight. The object of soupling is, indeed, to give to raw silk, if possible, all the properties of boiled-off silk, with the least loss of weight; considerable perfection has already been attained in this direction. Souple silk is not so strong as boiled-off silk, and is used for only tram.

The process of soupling consists essentially of two operations: first, the softening; and second, the soupling proper. With yellow silk, and whatever is intended to be dyed light colours, the operations of bleaching and stoving intervene.

(a) *Softening.*—The raw silk is worked for 1-2 hours in a solution of 10 % soap, heated to 25°-35° C. The object of this operation is to soften the fibre, and remove the small quantity of fatty matter present, so as to facilitate the operations which follow.

(b) *Bleaching.*—The silk is worked in stone troughs for 8-15 minutes in a dilute solution of aqua regia 4° Tw. (Sp. Gr. 1·02), heated to 20°-35° C. It is afterwards washed well till free from acid.

The aqua regia is prepared by mixing together five parts by weight of hydrochloric acid, 32° Tw. (Sp. Gr. 1·16), with one part of nitric acid, 62° Tw. (Sp. Gr. 1·31), and allowing the mixture to stand for 4-5 days at a temperature of about 25°-30° C. Before use it is diluted with fifteen volumes of water. For the aqua regia may be substituted sulphuric acid saturated with nitrous fumes, or a solution of the so-called "chamber-crystals," obtained in the manufacture of sulphuric acid.

The silk must not be worked too long in the acid liquid, otherwise the nitric acid causes it to contract a yellowish tint which cannot be removed. The moment the silk has acquired a greenish-grey colour it should be withdrawn from the bath, and well washed with cold water.

(c) *Stoving.*—This operation is similar to that already described. It renders the silk hard and brittle. Without removing the sulphurous acid, however, it is at once submitted to the following operation:—

(d) *Soupling* (Fr., *assouplissage*).—The silk is worked for about an hour and a half, at 90°-100° C., in water, containing in solution 3-4 g. cream of tartar per litre. The silk becomes softer and swells up, and being thus rendered more absorbent, it is better adapted for dyeing. The operation of soupling is a somewhat delicate one, and needs considerable judgment and practice. The solution must not be too hot, nor must the immersion of the silk be too prolonged, otherwise the loss in weight is excessive, and the result is unsatisfactory. After soupling the silk is finally worked in a bath of tepid water. Souple silk will bear warm acid baths subsequently, but not alkaline or soap baths beyond a temperature of 50°-60° C., otherwise it loses silk-glue, and is more or less spoiled. The operation of soupling is sometimes performed on raw silk which has been previously submitted to other operations, as, for example, in the dyeing of so-called black souples.

A satisfactory explanation of the theory of soupling has not yet been given. The cream of tartar probably acts as an acid salt merely, and although it gives the best results, it can be replaced by a solution of sodium or magnesium sulphate acidified with sulphuric acid, or even by very dilute hydrochloric acid. It is curious that silk is rendered less tenacious by soupling than by boiling-off.

Ecru Silk is raw silk which at most is submitted to

washing, with or without soap, and bleaching. The loss of weight varies from 1-6 %. Unbleached écru silk is only dyed in one or two different shades of black.

Bleaching Tussur Silk.—This may be accomplished according to the method proposed by M. Tessié du Motay, in which barium binoxide is the agent employed. Free baryta hydrate is first removed from the binoxide by washing the latter with cold water, and a bath is then prepared, containing binoxide in the proportion of 50-100 % of the weight of silk to be bleached.

The silk is washed for about an hour in the bath heated to 80° C., then washed and passed into dilute hydrochloric acid, and washed again. If the white is not good, the operations are repeated, or one may also complete the bleaching by washing the silk in a solution of potassium permanganate and magnesium sulphate, and afterwards in a solution of sodium bisulphite, to which hydrochloric acid has been added.

Although barium binoxide is little soluble in water, at the temperature of the bleaching bath it gives up oxygen to the fibre by degrees, even without the addition of any acid, so that the bleaching takes place gradually. During the immersion the silk also absorbs a certain amount of the binoxide, probably as hydrate, so that on the subsequent passage into acid there is liberated within the fibre hydrogen dioxide, which, being in the nascent state, bleaches to the best effect.

The chief difficulty of the process consists in the fact that, by long contact with the barium binoxide, the silk becomes dull, harsh, and tender, but with care the process can be made to yield excellent results and it is, indeed, already adopted in practice. Hypochlorite of ammonia has been employed, but with less success. The best bleaching agents for Tussur and also Mulberry silk are peroxide of hydrogen and peroxide of sodium, especially the former.

Tussur silk is bleached for the purpose of dyeing it in light colours, and a good bath is made up of a mixture of caustic soda, white curd soap, peroxide of hydrogen, and a little ammonia,

CHAPTER X.

WATER.

Soft and Hard Water.—Natural water occurs in the form of invisible vapour permeating the air. When the temperature becomes sufficiently low, the vapour condenses and becomes visible as dew, fog, or cloud, or is precipitated in the form of rain. The original source of natural water is the ocean. This is in a state of constant evaporation, and the vapour produced just as constantly undergoes the condensation alluded to. Indeed, a gigantic process of natural distillation is here presented, and, as one would anticipate, rain-water is the purest form of natural water.

A portion of the rain-water sinks into the earth until it reaches some impervious layer, from which it may be pumped up as well-water, or it flows underground, and eventually reappears on the surface as a spring, which frequently forms the source of a brook or river.

Another portion of the rain-water never penetrates the soil, but simply drains off the land, and forms the so-called surface-water, which also goes to form rivers.

The solvent power of water is so considerable that both spring and river water always contain certain mineral and vegetable matters, the nature of which varies according to the character of the rock or soil through or over which it has passed.

If the geological strata are composed of such hard insoluble rocks as granite and gneiss, the water remains comparatively free from impurities, and whether it flow as a river or rise as a spring, it will be what is termed a "soft" water.

If, on the contrary, the water during its subterranean course meets with rocks containing such a soluble constituent as rock-salt, it becomes brine; if it encounters a stratum of limestone, oolite, chalk, new red sandstone, etc., it dissolves a certain portion of it, and becomes magnesian or calcareous, and constitutes on its reappearance what is called a "hard" water. Common limestone being generally of a less permeable nature than magnesian limestone, does not yield such hard water as the latter. Again, if the water passes through or over rocks containing iron in some form or other, it takes up some of the iron and becomes a so-called "chalybeate" water. When the rain-water drains from boggy moorland, a certain portion of the more or less decomposing vegetable matter dissolves, and the water is usually brown-coloured.

The natural impurities of water which concern the dyer must be either suspended or dissolved, and of these the latter are the most important. A constant supply of clear water is certainly an indispensable requisite, but the means of purifying muddy water are comparatively simple, and the chemical nature of the suspended matter generally possesses little or no interest. The dissolved constituents of water, however, are equally, or even more, injurious, and to effect their removal is much more difficult, involving as it does the employment of chemical means of purification. As a general rule, river water contains the largest amount of suspended and vegetable matter, and the least amount of dissolved constituents, whereas spring and well water bear the opposite character.

Calcareous and Magnesian Impurities in Water.—These are at once the most frequently occurring, and the most injurious of all impurities. They are usually present as bicarbonates, less commonly as chlorides and sulphates. The latter are generally less injurious than the former.

The presence of lime is shown if the addition of a solution of ammonium oxalate to the water in question gives a white precipitate of calcium oxalate. On evaporating such a water to a small bulk it becomes turbid. If, then, the addition of hydrochloric acid produces effervescence, and renders the solution again perfectly clear, it is an indication that all the lime is present as bicarbonate. If there is no effervescence and no clearing, it is probably all present as sulphate. Effervescence and partial clearing denote the presence of both sulphate and carbonate. Should there be no turbidity on evaporation, lime is either absent altogether, or it is probably present as chloride or nitrate.

The presence of magnesia is detected, after lime and alumina have been removed by means of ammonia and ammonium oxalate. The filtered liquid is concentrated by evaporation and mixed with a solution of phosphate of soda and ammonia; magnesia is present if a white crystalline precipitate is thereby produced. The separation, however, of lime and magnesia has little or no importance for the dyer.

The presence of bicarbonates is further detected by the addition of a clear solution of lime-water producing a white precipitate.

Sulphates are present if an addition of hydrochloric acid and barium chloride gives a white precipitate.

A white curdy precipitate, which is produced on adding nitric acid and silver nitrate, denotes the presence of chlorides.

The injurious influence of magnesian and calcareous water cannot be overrated, and very specially because of its property of precipitating soap solutions. Hard water only produces a froth or lather with soap after the whole of the calcium and magnesium compounds present have been precipitated as insoluble lime and magnesia soaps, the latter of which may be distinguished from the former by their more objectionable curdy character.

By employing a soap solution of a standard strength, it is possible to calculate aproximately the amount of calcium and magnesium compounds present. Details of this method of determining the hardness of water (Clark's) are given in most text-books of chemical analysis.

It is well to note that, according to Clark's scale, a water with one degree of hardness contains 1 g. calcium carbonate per gal., but after the newer scale of Frankland, one degree of hardness signifies that the water contains 1 g. of $CaCO_3$ in 100,000 g. water. To reduce the degrees of the latter to those of Clark's scale it is simply necessary to multiply by seven-tenths.

In all those operations where large quantities of soap are employed, it is evident that the use of a hard water entails a considerable loss of soap. One part (by weight) of CaO is found to decompose about 15·5 parts of ordinary soap containing 30 % of moisture. After making a soap-analysis of the water, and knowing the quantity of the water employed, it is easy to calculate the annual loss of soap (about one-sixth) occasioned by the hardness of the water. Taking the monthly consumption of soap in London as 1,000,000 kg. = 948·2 tons, it is estimated that the hardness of the water used causes an expenditure of 230,000 kg. = 226·36 tons more soap per month than would be required if soft water were used.

This is, however, by no means the only disadvantage. The precipitated earthy soaps are more or less of a sticky nature, and adhere so tenaciously to the fibre that they cannot be removed by ordinary technical processes. In the scouring of wool or of silk they render the fibre more or less impermeable, so that neither mordant nor colouring matter can be afterwards properly fixed thereupon, and irregular development of colour results.

When the soap solutions are applied after dyeing—as in the clearing of Turkey-red, milling of woollen fabrics, etc.—the precipitated earthy soaps may impart to the finished fabric a pale greyish "bloom," or an unnatural lustre, and altogether ruin both the brilliancy of the colour and the value of the fabric.

In some cases earthy soaps may act injuriously by playing the rôle of mordants. It would be dangerous, for example, to employ soap in the bleaching of calico for printing purposes, or to re-dye printed calicoes after soaping, since any lime-soap precipitated on the fabric would attract colouring matter, and cause the white ground to be stained.

Hard water is also injurious in the dye-bath, because it only imperfectly extracts the colouring matter from the dye woods employed. Some colouring matters—Alizarin Blue, Cœruleïn, etc., also Catechu and Tannin matters—produce insoluble compounds with the alkaline earths, and may thus be precipitated and rendered inactive.

It must not be forgotten, however, that the presence of a certain limited amount of lime is beneficial, and, even necessary, in dyeing with some colouring matters—as Alizarin, Logwood, Weld, etc.—but in such cases even, it is always preferable to have a pure water, so that one may add the most suitable form and amount of lime salt.

Hard water has generally the effect of dulling the colours obtained from many colouring matters, both during the dyeing and the subsequent washing processes. When the hardness is due to the presence of earthy bicarbonates, it retards or even prevents the dyeing of such colours as are produced only in an acid bath, as cochineal scarlet.

Such water acts injuriously also on solutions of certain mordants—like those of aluminium, iron, etc.—by neutralising a portion of their acid and precipitating basic salts, the mordanting bath being then less effective.

In some cases of mordanting, however, water containing earthy bicarbonates is to be preferred to pure water, since it fixes a much larger quantity of insoluble basic salt on the fibre, as, for example, in the washing of silk after mordanting with basic ferric sulphate, and with aluminium or tin mordants.

Water rich in earthy bicarbonates is not suitable for the solution of many of the coal-tar colours, such as Methyl Violet, etc. A portion of the colour-base is precipitated as a tarry mass, and not only is colouring matter wasted, but goods dyed in such solutions are apt to be spotted.

Ferruginous Impurities in Water.—These are to be looked for in water which is derived from disused coal-pits, iron-mines, iron and aluminous shales, etc., and are also very objectionable in dyeing operations. Surface water draining off moorland districts and passing over ochre-beds also contains iron, evidence of which is seen in the brown ferric oxide deposited on the stones in the stream. All such water should be rigorously avoided.

To test for the presence of iron, evaporate some of the water in a clean porcelain basin. If a reddish-brown deposit is thereby produced, this must be collected, dissolved in a little hydrochloric acid, and thoroughly oxidised by heating, with the addition of a little potassium chlorate. The solution is diluted and cooled, and a solution of potassium ferrocyanide, or thiocyanate, is added. If iron is present, a blue precipitate or red coloration, respectively, is thereby produced.

The iron being usually present as bicarbonate, acts upon soap solutions after the manner of the analogous calcium and magnesium compounds, and similar or even worse results ensue.

In wool-scouring, cotton-bleaching, and other operations where alkaline carbonates are used, ferric oxide is precipitated upon the fibre. With such goods it would be quite impossible to dye bright colours subsequently— e.g. alizarin reds, etc.—and all colours, indeed, suffer more or less. Bleached fabrics acquire an unpleasant yellowish tinge, which will neither rinse nor scour out, and which renders the goods quite unsaleable.

Alkaline Carbonates as Impurities in Water.—Water containing sodium carbonate is frequently met with in districts where the supply is derived from wells which penetrate the lower beds of the Coal Measures. This alkaline condition is detected by means of red litmus-paper.

Such water is by no means detrimental in wool scouring, or in operations where alkaline carbonates are nominal constituents of the bath, if other injurious constituents are absent; but for purposes of mordanting, dyeing, and washing of dyed goods, it is, if possible, more injurious than the water containing earthy carbonates. When no other water is to be had, it must for such operations be carefully neutralised with sulphuric or acetic acid.

Acid Salts and Free Acids as Impurities in Water.— Water draining from moorland districts contains what are usually termed peaty acids, and since these attack iron very readily, they may indirectly cause serious injury.

Water derived from shale beds containing pyrites and situated near the surface, becomes contaminated with ferrous sulphate. On exposure to air this salt oxidises, ferric oxide is deposited, and the water contains free sulphuric acid. Blue litmus-paper serves to detect the presence of this impurity.

Such water is unsuitable for scouring operations, since it decomposes and wastes the detergents used. If soap is employed, fatty acid is liberated and liable to be fixed upon the fibre. In dyeing operations acid water is equally injurious. If such water cannot be avoided, it must be carefully neutralised with carbonate of soda.

Organic Impurities in Water.—These as they exist in moorland streams have not been found in practice to be injurious, unless the water is thereby so deeply coloured as to stain the woollen or other fibre, which is to be dyed light shades only, or is intended to remain in the bleached condition. In the absence of inorganic reducing bodies the presence of organic matter is determined by adding a solution of permanganate of potash sufficient to impart a pink colour, acidifying with sulphuric acid, and then boiling. If organic matter is present the solution is decolorised.

Sulphuretted Hydrogen as an Impurity in Water.—This is only occasionally met with, and arises through the decomposition of gypsum by organic matter. Such water is to be rejected, since, when iron, tin, copper, and lead mordants are used, sulphides of these metals are formed, and cause black or brown stains.

This impurity is detected by slightly acidifying the water with acetic acid and placing it in closed flasks in a warm place for some time. A strip of filter paper moistened with lead acetate is fixed above the surface of the liquid. The formation of brown lead sulphide on the filter paper denotes the presence of sulphuretted hydrogen.

Specially injurious are the impurities which may come from establishments situated higher up the stream—*e.g.* paper-works, chemical-works, bleach-works, etc. Hence the dyer, whose only source of supply is a river running through a manufacturing district must be specially vigilant, for although the pollution of rivers by waste products, etc., from various manufactories may be forbidden, the compliance with this restriction is beset in many instances with enormous difficulties which, indeed, can only be partially overcome.

Correction and Purification of Water.—Pure water is certainly one of the first requisites for all operations of bleaching and dyeing. Unfortunately it is at the disposal of few manufacturers, and the increasing injury thus caused has not always met with the attention it deserves. Directors of bleach and dye works will find this subject as difficult as it is interesting and important.

Although it is comparatively easy to purify, or at least to correct, small quantities of water, the technical problem of readily purifying such large supplies of water as are necessary for dyeing, etc., is surrounded with difficulties. The purification of water usually comprises both a mechanical and a chemical treatment.

Mechanical Purification of Water.—Whenever the land is suitable, the water is collected and allowed to settle in large reservoirs, even if only for the purpose of storage. When possible, several reservoirs are maintained, and it is well to pass the water from one to the other in such a manner as to expose it as much as possible to the air— for instance, by a staircase cascade. Finally, the water should be passed through filter-beds of sand. During the exposure to air in this manner a partial chemical purification will take place in water containing calcium, magnesium, or iron bicarbonates; a portion of the solvent carbonic acid is lost and the carbonates are precipitated.

The most favourable circumstance is that in which the works are placed upon the bank of a river flowing from a lake situated immediately above, so that neither heavy rains nor melting snow will render the water turbid.

Purification of Water by Boiling.—Reference has already been made to the convenient classification of "soft" and "hard" waters. The latter may be sub-divided into those which are temporarily hard, and those which are permanently hard, although most waters combine both properties.

A water possessing temporary hardness becomes soft by mere boiling, if this is sufficiently prolonged, since such hardness is due to magnesium, calcium, or iron bicarbonates. The effect of boiling is to expel one-half of the carbonic acid, and thus to precipitate the insoluble monocarbonates produced. Since the softening effect takes place only gradually, the boiling should be continued for 20-30 minutes at least. It is scarcely necessary to add that this method of purification is far too costly to serve for large quantities of water.

A water which is permanently hard derives this property from the presence of sulphates of the above-mentioned metals; hence, in this case, boiling has no softening influence; on the contrary, the hardness is increased through the concentration of the water, and other modes of purification must be adopted.

Chemical Purification of Water.—As to the purification or correction of water by chemical means, one element of difficulty is the inconstancy of the composition of the water. Heavy rain and melting snow dilute it, while hot summer weather concentrates it, especially in small rivers. Whenever it is impossible to keep pace with varying conditions of this kind by making frequent analyses of the water, it is advisable, in the absence of purification or correction on a large scale, to add to the water such agents as are not specially injurious, even if added in slight excess. In many cases it suffices to neutralise as carefully as possible the alkalinity of a water arising from the presence of earthy bicarbonates by the addition of acetic acid—*e.g.* in most cases of dyeing, or for the purpose of dissolving coal-tar colours. An old method of purifying small quantities of water which is still often used by silk and woollen scourers, etc., is to boil the water with the addition of a little soap, with or without the addition of sodium carbonate, and to skim off the earthy soaps thrown to the surface. Apart from its expense, this method is unsatisfactory, since the major portion of the earthy soaps, etc., remain disseminated in the water in a finely-divided state. Water corrected in this manner is necessarily left in an alkaline condition from the alkali of the soap remaining behind, and of course the amount of alkaline carbonate left is equivalent to that of the earthy salts removed, thus :—

$$\underset{\text{Calcium bicarbonate.}}{CaH_2(CO_3)_2} + \underset{\text{Soap.}}{2C_{18}H_{33}O_2Na} = \underset{\text{Lime-soap.}}{Ca(C_{18}H_{33}O_2)_2} +$$
$$\underset{\text{Sodium carbonate.}}{Na_2CO_3} + CO_2 + \underset{\text{Water.}}{H_2O}.$$

The dyer's method of adding alum to the water of the mordanting or dye-bath, and then boiling and skimming off impurities which rise to the surface, is even more uncertain, and is strongly to be deprecated.

Purification of Water with Lime. Clark's Process.—Since calcium and magnesium carbonates are soluble in water only by the presence of carbonic acid, the natural remedy is to employ some means which will rapidly and effectively remove or absorb the latter. In 1841 Dr. Clark proposed calcium hydrate or slaked lime as the cheapest and most suitable agent for this purpose, and his method, or some modification of it, is still generally adopted. The following equation explains the theory of the process :—

$$CaH_2(CO_3)_2 + Ca(OH)_2 = 2CaCO_3 + 2H_2O.$$
Calcium bicarbonate.   Calcium hydrate.   Calcium carbonate.

In carrying out the method here expressed, many practical difficulties are met with, and constant skilled oversight is necessary to insure success.

Only the temporary hardness is removed, and not even this completely. A small residuum of chalk always remains in solution. It is quite possible, however, to remove 10-11ths of the whole temporary hardness, as well as iron salts and much organic matter. Experiments on a large scale have proved that, by the lime process, water of 23° hardness can be reduced to 7°, of 15° to 3° or 4°, and so on.

It is best to employ clear lime-water for correction, since this possesses a known constant composition. The amount of such lime-water which it is necessary to employ may be calculated after making an analysis of the water, or may be determined by actual experiment. If the number of degrees hardness (Clark's scale) is divided into 130 or 150, the number obtained will approximately represent, as a rule, the number of litres of water which can be softened by the addition of one litre of lime-water.

Clear lime-water may be replaced by milk-of-lime, if the latter is carefully applied. Excess of lime in the corrected water is readily detected by adding a little of it to a filtered decoction of cochineal. Such excess changes the yellowish-red colour of the solution to a violet. Other delicate alkali-indicators may also be adopted.

In working Clark's process as originally devised, large tanks or reservoirs are requisite. It is best to have at least three—one into which to run the water and lime and to allow the precipitated chalk to settle for about 16 hours; another from which clear and previously corrected water can be drawn; and a third as a reserve during cleansing operations. Each reservoir should hold at least a day's supply.

Purification of Water with Caustic Soda.—For purposes of scouring, or where a slightly alkaline water is not prejudicial, caustic soda may be conveniently substituted for quicklime, since its solution can so readily be made of a standard strength and added in the requisite amount to the water to be corrected:—

$$CaH_2(CO_3)_2 + 2NaHO = CaCO_3 + Na_2CO_3 + 2H_2O.$$

In this case, of course, exactly the same amount of

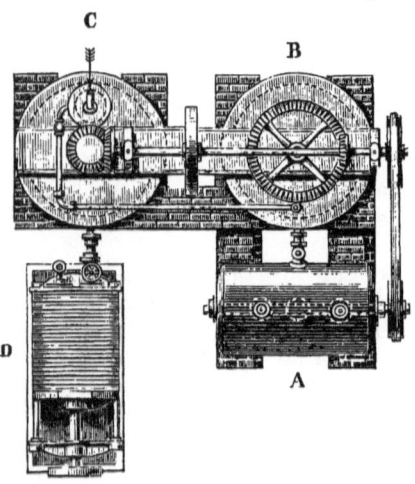

Fig. 39.—Porter-Clark's Apparatus for Softening Water—Plan.

sodium carbonate remains in the corrected water as if the purifying agent used had been soap. It is well to heat the mixture to 50° C. in order to cause more rapid settling of the precipitate. Mechanical impurities, also iron, aluminum, and earthy phosphates are completely thrown down.

Permanent as well as temporary hardness is removed by the use of caustic soda, since the calcium and magnesium sulphates present are decomposed by the sodium carbonate produced in the above reaction.

Should the water corrected in this manner be required

for purposes where alkalinity is to be avoided, it can be readily neutralised before use.

Porter-Clark Process of Softening Water.—The essential improvement effected by this process is a saving of space, time, and labour, through the application of machinery to the ordinary Clark's process.

Figs. 39 and 40 give plan and elevation of an arrangement for supplying over 6,000 litres = 1,320·58 gals. of

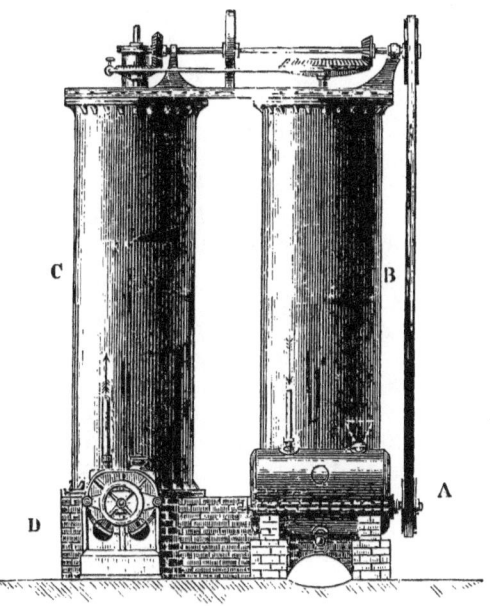

Fig. 40.—Porter-Clark's Apparatus for Softening Water—Elevation.

softened water per hour, but there is practically no limit to the quantity which may be supplied if the apparatus is made large enough.

The lime-water is prepared in the small horizontal cylinder A by constantly churning up slaked lime with water admitted under pressure direct from the reservoir or main. By a pipe midway in the height of the churn, the more or less saturated lime-water (a saturated solution contains about 1·4 g. of lime per litre = 0·22 oz. per gal.), and with some lime in suspension, is led into the large cylindrical vessel B, where the lime and water are kept in slight agitation to assist in completing the saturation. As the lime-water ascends, the particles of lime in suspension gradually settle out, and tolerably clear lime-water passes out at the top into the cylinder C, where it is continuously mixed with the water to be purified in accurately determined proportions. The supply of each is regulated by valves furnished with dial plate and index. A brisk agitation is maintained in the mixing cylinder C, in order to facilitate the chemical reaction taking place. When this is completed, the chalky water is forced through the filter-press D, wherein the carbonate of lime acts as a medium of filtration, and the clear water thus obtained is at once fit for use.

If it be desired to remove both permanent and temporary hardness by means of the above apparatus, carbonate of soda, or caustic soda, must be used in addition to the lime.

Gaillet and Huet's Process of Softening Water.—The agents of purification here adopted are lime and caustic soda, and the cost does not exceed, say, one farthing per 1,000 litres = 220·09 gals. of softened water.

This apparatus differs essentially from all others by the simple but effective means adopted for separating and removing the precipitated impurities without in any way choking the purifier or retarding the delivery of water.

Fig. 41 gives a perspective view of a complete apparatus capable of purifying about 200,000 litres = 44,000 gals. of water per day. The purification takes place in the large square clarifying, or precipitating, tank A, which forms the body of the apparatus, and which is shown separately in Fig. 42, a portion of the casing being there removed to show the interior arrangement. Instead of passing downwards through filtering screens, which would soon become clogged and retard the flow, the water, after having been mixed with the precipitating reagents, enters at the bottom of the tank, A, and rises slowly towards the top, following a ziz-zag course in the shallow spaces between a number of V-shaped diaphragms, inclined at an angle of 45°, and riveted alternately to opposite faces of the tank, as also to the two adjacent sides. All the diaphragms shelve at the same angle towards the same face of the tank, where they lead to a series of mud cocks, F.

Above the clarifying tanks are situated smaller tanks, in which the precipitating reagents are dissolved. Tank B contains the solution of caustic soda, of which the required quantity is run off into one of the tanks, C, in which lime has been dissolved in water. The liquid is allowed the requisite length of time to settle, the other tank, C, being meanwhile in use. The clear soda-lime solution thus obtained is run off at a regulated rate, and mixes with the water to be purified entering at D, in a special tank situated above the clarifying tank, A, and immediately below the raised platform. The turbid water falls through the pipe E, enters the clarifying tank at the bottom, and at once assumes an upward motion. As the section of the tank A is very large, and that of the supply pipe D very small, the water naturally rises very slowly and gently, allowing the precipitate to settle almost as if the water were at rest. The purpose of the V-shaped diaphragms will now be apparent. The water has to pass slowly between them in shallow layers, and as the solid particles have but a few inches to fall, they readily settle upon the diaphragms, which represent, indeed, a very large precipitating area in a limited space. As the latter are V-shaped, the deposit slides down into the angle towards the mud cocks, F, through which it is discharged when necessary. The water meanwhile becomes gradually clearer as it rises, and is ultimately drawn off at the top, G, perfectly soft and limpid.

The process of purification is carried on automatically without the necessity of constant attendance or motive power. It need scarcely be added that the amount of soda-lime solution to be mixed with the water is determined according to the results of a careful analysis previously made of the water.

Other modifications of Clark's process are in vogue, differing chiefly by the mode of effecting the clarification of the water after mixing with the precipitating reagents, but the two processes described may be considered typical of the rest.

Purification of Water Discharged from Dye-houses.—
If it is necessary that the dyer should have pure water

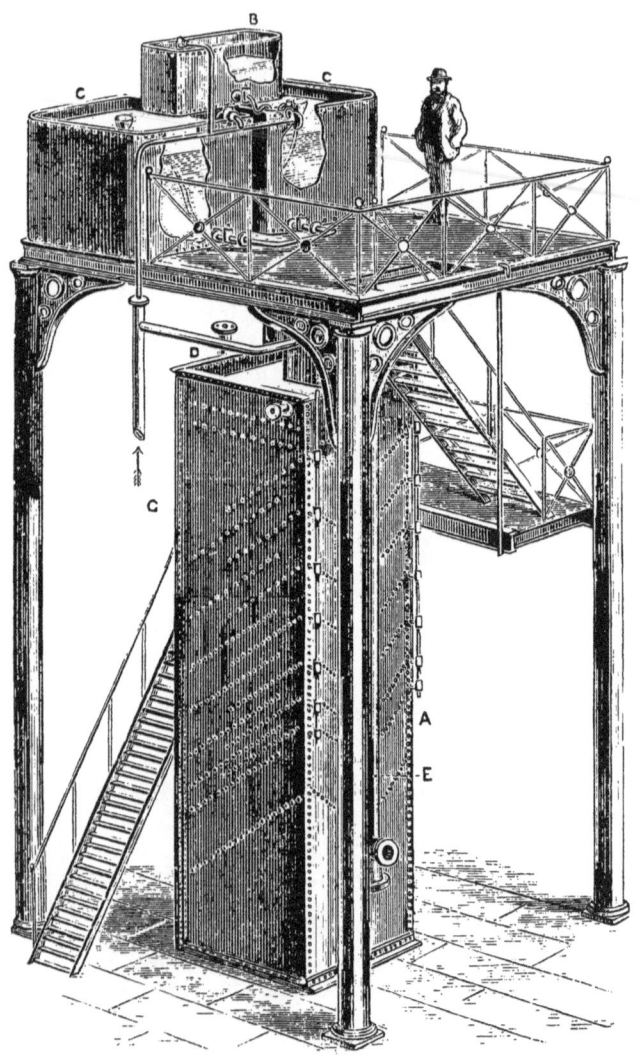

Fig. 41.—Gaillet and Huet's Apparatus for Softening Water.

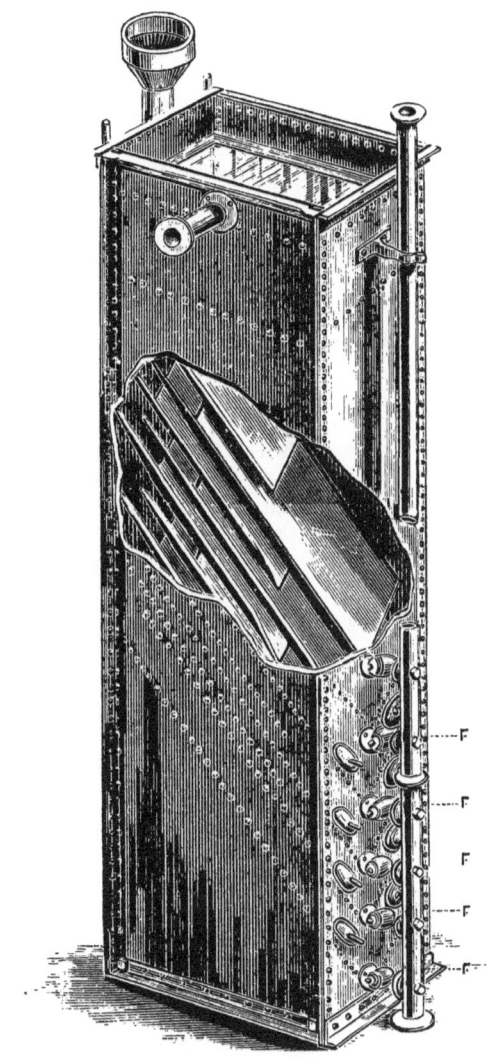

Fig. 42.—Gaillet and Huet's Precipitating Tank.

for the successful prosecution of his business, he ought to feel it his duty not to pollute the river, from which he possibly receives his supply, with injurious discharges, to the annoyance and loss of his less-favoured neighbours lower down the stream.

Not only are the waste liquors from dye-works for the most part highly coloured, but they contain large quantities (over 1·5 g. per litre) of organic and inorganic matter, both suspended and dissolved. Should the dye-works situated on the banks of a small stream be numerous, the latter generally assumes a turbid, inky appearance; its bed is gradually impregnated with decomposing organic matter, putrescent odours are given off, the water is poisoned, and the stream becomes practically a large open drain, disagreeable to the sight and more or less noxious to health.

One of the simplest modes of mitigating this evil is to conduct all the refuse waters resulting from the various operations of the works into two or more reservoirs, where they mix together and precipitate each other. The whole must be allowed ample time to settle, and only the clear water permitted to flow into the river. Where space is limited, filtering beds of coke, sand, etc., may take the place of settling reservoirs. The purification is rendered more complete, however, if additional precipitating agents are employed, such as magnesium and calcium chloride, lime, etc. Of these lime is perhaps the cheapest and the most generally efficacious; it neutralises acids, precipitates colouring matters, mordants, soapy liquids, albuminous matter, etc.

That such a simple means can accomplish the end in view in a most satisfactory manner is exhibited in the large works of Mr. W. Spindler, at Cöpenick, near Berlin, in which are carried on all branches of dyeing, printing, and finishing of silk, woollen, and cotton goods.

In this establishment, according to Caspari, all the refuse water flows into two large collecting reservoirs, where the suspended matter is allowed to settle. The supernatant water is shown by analysis to be strongly impregnated with salts of the alkalies and iron, together with tannin and extractive matter from dyewoods, fatty matter, colouring matter, etc. After being transferred to another reservoir, therefore, it is mixed with lime-water and a solution of calcium chloride; precipitation ensues, and the whole turbid mixture is pumped into still larger reservoirs, where it is allowed to settle. The clear water now obtained is found to be simply a hard water containing a small amount of organic matter, and is either used for purposes of irrigation or allowed to drain through the soil into the river.

The solid matter which has settled in the first collecing reservoirs is dried and calcined in gas retorts, and yields per 100 kg. 13-16 cub. mètres of illuminating gas (per 1 cwt., 232-285 cub. ft.).

In bleach-works the refuse liquids consist of alkaline and soapy solutions, together with such as contain calcium chloride, traces of bleach-powder, and free acids. Here are all the elements necessary to mutual purification, if allowed to mix together in due order and proportion; the calcium chloride will precipitate the soapy solutions, while the free acids will neutralise and precipitate the alkaline liquids and decompose the waste solutions of bleaching-powder.

It is impossible that any single method of purification should be applicable to all works, but the following description of a satisfactory process, devised by Messrs. R. and A. Sanderson and Co., of Galashiels, and adopted by all the woollen manufacturers of that town, will be of interest, and may indicate what ought to be done in this matter in woollen mills.

The effluent water from woollen dye-works consists partly of solid matter in suspension—such as spent dyewoods, fragments of woollen fibres, etc.—and partly of soluble substances contained in the waste dye, mordant, and scouring liquors. The method of purification is, of necessity, therefore, chemical as well as mechanical.

Fig. 43 gives a plan of the purification plant in use at Messrs. Sanderson's works, where upwards of 40,000 litres = 8,800 gals. of waste scouring-liquor and about 120,000 litres = 26,400 gals. of dye and acid magma water are treated daily.

In order to reduce as much as possible the amount of waste water to be purified, the soapy liquids from the yarn and piece scouring are used again in the wool-scouring. The refuse liquid from this operation is run through a sieve into the settling tank, A, whence it overflows at a given point into the large reservoir B. From here it is pumped into the high level magma tanks, C (each of about 20,000 litres = 4,400 gals. capacity), where it is thoroughly well mixed, by means of an air-pump, P, with a calculated quantity of sulphuric acid. After being allowed to settle, the supernatant acid liquid is run off into the tank E for further treatment. The precipitated magma of fatty matter is allowed to flow into the pit D, the bottom of which is a drainer made up of ashes, spent dyewood chips, and sawdust. Here it is allowed to drain

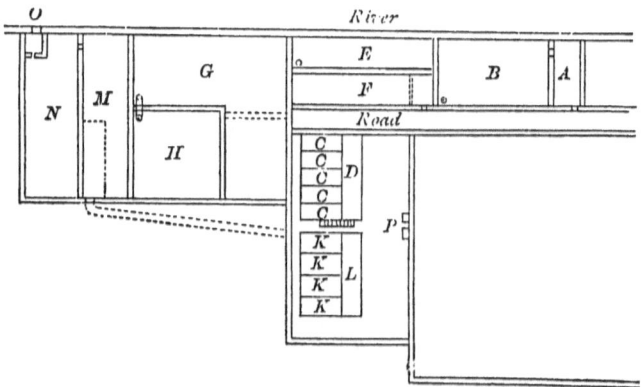

Fig. 43.—Plan of Purification Works for Waste Dye-liquors.

for about a week, the acid filtrate being also led into the tank E. The pasty magma is sold to oil-extractors and soap-makers.

The highly-coloured discharge from the dye-house passes through a sieve into the settling tank F, whence it overflows into the large reservoir G. In this reservoir the colouring matter of the spent dye-liquor, and the unexhausted mordants, partially combine and precipitate each other. Tank H contains slaked lime, well mixed with some dye-house liquor from tank G.

By means of the pump P definite proportions of acid magma water, from E (about one part), spent dye-liquor, from G (about three parts), and lime-water from H (about one part), are forced into the high-level purification tanks, K, and there thoroughly well mixed. By this means the whole of the colouring matter is thrown down as a fine, flocculent precipitate. After being allowed to settle, the almost colourless supernatant water flows into the tank M, and the deposit is run off at intervals into the pit L, and there allowed to drain until it acquires such a consistency that it can be dug out and conveyed to the refuse heap. The water from the purification tanks K is slightly alkaline from excess of lime, but in the tank M it is neutralised by allowing it to mix with the slightly acid rinsing-water coming from the dye-house. Thus purified, the water overflows into the tank N and passes through a filter into the river O, clear, neutral, and free from objectionable colour.

The following is an analysis of the effluent water by Crum Brown :—

Total solids in grams per litre.		Turbidity	Colour.	Ammonia in grams per litre.		Analysis of inorganic solids in grams per litre.	
Organic	Inorganic			Saline	Albuminoid		
0·13	0·78	1	3	0·003	0·002	$K_2Cr_2O_7$	0·006
						$CaSO_4$	0·11
						$CaCO_3$	0·32
						Na_2SO_4	0·29
						$NaCl$	0·06

The turbidity is represented by the quantity of china clay (stated in g. per 70 litres), which, when added to pure water, gives a turbidity equal to that of the sample.

The colour is represented by the quantity of ammonia (stated in hundredths of a g. per 70 litres), which, when added to pure water, gives, with Nessler's reagent, a colour as nearly as possible agreeing with that of the sample. The colour of the effluent water is a faint brownish-yellow.

This analysis shows that, with the exception of the small quantity of potassium bichromate, the substances remaining in solution in the water discharged into the river, are the usual substances found in ordinary spring and river water.

In explanation of the above process, it may be stated

that part of the lime used serves the purpose of neutralising the acid magma liquor, while the excess precipitates, from the waste mordant liquor, the hydrated metallic oxides, which at once combine with and precipitate any colouring matter present. It will be observed that, as far as possible, the waste materials from the various processes of manufacture are caused to aid in their mutual purification and removal from the water discharged into the river.

In the works of E. Schwamborn at Aachen, the refuse water from the washing of raw wool, and the milling and washing of cloth, is precipitated by lime. The composition of the air-dried precipitate is as follows:—

Water ...	3·11%
Lime and ferric oxide	18·47%
Fatty matter	71·96%
Wool fibre, &c.	6·46%

Mixed with coal, this precipitate serves for the manufacture of illuminating gas. Although in this method the potash salts of the raw wool are lost, it is estimated that, after deducting the working expenses, there is a net recovery of 30 % of the value of the soap used in milling. In other works the precipitate is treated for the recovery of fat.

CHAPTER XI.

ABOUT DYEING.

Materials and Colouring Matters Used by Dyers.—The two most essential elements with which the dyer has to deal are the material to be dyed and the colouring matter to be applied. With regard to the former, our attention is confined to the textile fibres, and it has already been shown that great differences exist between them, both as to their physical and their chemical properties. Not unnaturally, therefore, they might be expected to behave differently towards colouring matters. That such is really the case is readily shown by making the following simple dyeing experiment.

Three pieces of clean white textile material—wool, silk, and cotton—are immersed in a moderately strong aqueous solution of acid Indigo Extract, and are kept in continual movement by stirring, while the liquid is gradually heated to the boiling point. If the pieces are then taken out and well washed with water they present a remarkably different aspect; the wool and silk have become dyed, and appear pale or deep blue according to the amount of colouring matter employed, while the cotton is not dyed, or, at most, becomes slightly stained. If the experiment is repeated with many other colouring matters —such as Magenta, Methyl, Violet, etc.—similar differences are noticed.

The real cause of this striking difference in behaviour towards colouring matters exhibited by the different textile fibres is still a matter of discussion. Several theories have been propounded, but none has gained general acceptance.

It is maintained by some that the animal fibres attract certain colouring matters by reason of chemical affinity, and that in the process of dyeing they actually combine with the soluble colouring matter to produce an insoluble coloured compound. Cotton, they say, does not become

J

dyed because it has no affinity for such colouring matters. Such is the purely chemical theory of dyeing.

The opponents of this theory very properly urge as a vital objection to it that two fundamental signs of chemical combination having taken place are entirely wanting, namely, the union of the fibre and colouring matter according to chemical equivalents, and the disappearance of the special properties of each.

Experiment proves that the animal fibres may attract either large or small amounts of Indigo Extract, and be properly dyed in both cases, the difference being simply one of intensity of colour. There is not the slightest appearance of a combination according to molecular proportions, as in the formation of Prussian Blue by the mutual reaction of ferric chloride and yellow prussiate of potash.

As to the second point of the objection, no doubt the soluble Indigo Extract attracted by the fibre has apparently become more or less insoluble in water, but it can be readily removed—by heating with a dilute solution of carbonate of soda—with all its properties unchanged, and the decolorised fibre is also exactly the same as before.

Those who adhere to the mechanical theory of dyeing explain the foregoing facts, by stating that the animal fibres have become dyed by reason of a purely physical attraction exerted between the Indigo Extract and the said fibres. The latter, they say, are very porous or absorbent bodies, the colouring matter has penetrated the substance of the fibre, and is there retained in an unchanged condition.

By this theory the whole question is resolved into one of surface attraction, and the action is said to be identical with that which takes place when a weak solution of the same colouring matter is decolorised by filtering through animal charcoal.

On the whole, the mechanical theory of dyeing seems to have most points in its favour, although it cannot be denied that an alteration in the chemical composition of a fibre may materially alter its behaviour towards certain colouring matters. The most striking example of this kind is that exhibited by cotton, when changed into oxycellulose through the action of hyochlorous acid.

Pigments and Colouring Principles in Dyeing.—To

return now to the second of the two essential elements dealt with by the dyer, namely the colouring matter; so excessively varied are the bodies belonging to this class, both in physical and chemical properties, that it is again not towards the same fibre.
at all surprising that they should behave differently, even

If two pieces of wool are treated in separate vessels, the one with a solution of Magenta and the other with Alizarin, the former will soon be dyed red, while the latter only assumes a brownish-yellow stain of no practical use.

If a third piece of wool is first heated in a solution containing a suitable amount of aluminium sulphate and cream of tartar, and, after washing well, is then boiled with Alizarin and water (preferably somewhat calcareous), it acquires a bright red colour.

When other metallic salts—as potassium dichromate, stannous chloride, ferrous sulphate, etc.—are substituted for the aluminium sulphate, the wool becomes dyed other colours, namely, claret-brown, orange, purple, etc.

If similar experiments are made with magenta, the wool always assumes a more or less similar magenta-red tint.

It is quite evident from these experiments that Magenta and Alizarin have totally different dyeing properties; they are, indeed, typical representatives of two distinct classes of colouring matters.

The members of the one class are coloured bodies or pigments in which the colour is fully developed, and they require simply to be fixed on to the textile fibre, more or less, in their unchanged state, to cause the latter to become dyed. Such colouring matters may be conveniently termed monogenetic, since they are only capable of yielding, at most, various shades of one colour. To this class belong Magenta, Indigo, Orceïn (orchil), Picric Acid, Methyl Green, etc. The members of this class may be soluble (Magenta), or insoluble (Aniline Black), organic (Indigo), or inorganic (Ultramarine Blue).

As to the members of the other class of colouring matters, although they are generally possessed of some colour, this is not an essential feature, and even when present it generally lacks intensity, and does not necessarily bear the slightest relationship to the colours obtainable from them in dyeing. As a rule, they are to be considered as colouring principles capable of yielding several colours, *i.e.* very distinct coloured bodies, according to the means employed for the production of the latter, and hence an appropriate name for them may be polygenetic colouring matters. To this class belong Hæmateïn (logwood), Alizarin (madder), Galleïn, etc.

From what has just been stated, it will be understood that the general mode of applying the members of these two classes of colouring matters in dyeing is very different. This is, however, only partially true, for the distinction between monogenetic and polygenetic colouring matters in this respect is not sharply defined, since there are those which stand, as it were, on the borderland between the two. Some of the monogenetic colouring matters—as Alizarin Blue, Cœruleïn, etc.—combine the properties of colouring principles and of veritable pigments. The method of applying them to the textile fibres is that usual with the polygenetic class, but they are only capable of yielding various tones of one colour, and they can also be applied by special methods adopted with certain monogenetic colouring matters, for instance, Indigo.

The presence of the fibre in the third experiment cited above is not a condition essential to the development of the colour, as can be readily enough shown by the following experiment: Make a dilute solution of aluminium sulphate, and render it somewhat basic and more sensitive by neutralising a portion of its sulphuric acid with sodium carbonate; add now to the still clear solution a little Alizarin, and shake the mixture vigorously, or heat it a little. A red-coloured body is very soon produced in the form of an insoluble precipitate, especially if a calcium salt be also present. It would appear in this case that the colouring matter enters into chemical combination with the aluminium, or with a very basic salt of the same.

Analogous but variously-coloured precipitates are produced on substituting decoctions of Cochineal, Persian berries, Logwood, etc., for the Alizarin, or by replacing the aluminium sulphate with solutions of other metallic salts. Coloured precipitates produced in this manner are the real colours or pigments which it is desired to obtain from the polygenetic colouring matters; indeed, when dried, ground, and mixed, say, with boiled oil, they are used by the painter under the name of *lakes*. The object of the dyer, however, is not only to produce, but, at the same time, to fix these coloured precipitates or lakes on the material to be dyed, without the aid of such a vehicle as boiled oil.

To effect this, two operations as a rule are necessary, namely, mordanting and dyeing.

The Mordanting Process.—The first has for its object the precipitation and fixing upon the textile material as firmly and permanently as possible, of some substance capable of combining with the colouring matter subsequently to be applied, and precipitating it in an insoluble state upon the fibre. This operation, or series of operations, constitutes the mordanting process, and the metallic salts or other substances used for this purpose are termed mordants. This name is derived from the Latin, *mordere*, to bite. It was originally introduced because the early French dyers considered that the utility of these metallic salts consisted in their corrosive nature; the general opinion was that they simply made the textile fibres rough, that they opened their pores, and thus rendered them more suitable for the entrance of the colouring matters. No doubt some slight corrosive action does take place here and there, in which case, of course, the surface of the fibres will be increased, entailing, probably, a slight increase of physical attraction of the fibre for colouring matters, but the essential action of mordants is undoubtedly chemical. As to the manner of applying these mordants, it varies with the different origin and state of manufacture of the textile fibres, the nature of the mordants and colouring matters employed, the particular effects to be obtained, and so on.

The method generally adopted in the case of the woollen fibre is to boil it with dilute solutions of the metallic salts, frequently with the addition of acid salts—such as cream of tartar, etc. A partial dissociation of the metallic salts takes place, induced and augmented both by the dilution and heating of the solution and the addition of assistant acids and salts, and partly by the presence of the fibre itself. What actually becomes fixed upon the fibre during this process is, in most cases, more or less a matter of conjecture, the chemistry of the process having been as yet only imperfectly studied, although recent research has been turned in that direction.

Although the term mordant is generally applied only to the metallic salts used, a mordant in the widest sense of the term is that body, whatever it may be, which is fixed on the fibre in combination with any given colouring matter. In the case cited of dyeing wool Alizarin red, it is considered that during the boiling of the wool with cream of tartar and aluminium sulphate, this latter salt is decomposed, and the mordant precipitated on the fibre is insoluble alumina, or a basic aluminium sulphate; in the subsequent dye-bath this combines with the Alizarin to produce the red-coloured compound, or lake.

Silk and wool have many analogous properties, and hence the silk fibre is frequently mordanted in a manner similar to that employed for wool, but as a rule high temperatures are avoided, and mere immersion in a cold concentrated metallic salt solution, with a subsequent washing with water, is all that is necessary. During the immersion the silk absorbs the metallic salt more or less unchanged, but in the subsequent washing this absorbed salt is dissociated by the mere dilution with water, and an insoluble basic salt is precipitated within the substance of the fibre. Certain soluble basic ferric sulphates, basic aluminium sulphates, and stannic chloride, behave in this manner.

As to the methods employed for mordanting cotton, they are usually more complex, since this fibre has not the property, like silk and wool, of decomposing metallic salts by such a simple method as boiling in their solutions, nor is it so porous as these fibres. In some cases metallic salts are chosen whose component parts are, under certain conditions, readily separable from each other, such as acetates of iron and aluminium, and with these, after immersing the cotton in their solutions, and removing the excess, it may suffice to dry the cotton and then to expose it for some time to air rendered suitably warm and moist ("ageing.")

Ferrous acetate absorbs oxygen from the air most energetically, and forms already in this way a basic salt soluble with difficulty.

$$2[Fe(C_2H_3O_2)_2] + O + H_2O = Fe_2(C_2H_3O_2)_4(OH)_2$$
Ferrous acetate.   Basic ferric acetate.

In most cases of mordanting, however, whether the material be silk, wool, or cotton, there is precipitated upon the fibre a metallic oxide, or a basic metallic salt, with a corresponding liberation of acid, or formation of an acid salt, and the mordanting process continues only as long as the constantly increasing acid allows, *i.e.* until a state of equilibrium proper to the new conditions has become established.

The advantage of using the metallic acetates as mordants for cotton is that the liberated acid does not injure or tender the fibre, which is readily the case with other salts; further, owing to its volatility, the acetic acid is quickly removed, under suitable conditions of heat and moisture, from the field of action, and a more perfect precipitation of the real mordanting body on the fibre takes place.

In some cases of mordanting cotton the use of acetates and the simple exposure or "ageing" referred to are avoided, either from motives of economy or because certain practical difficulties arise. With cotton yarn, for example, the drying is apt not to be uniform, and irregular colours result from unequal decomposition of the acetate. Even where acetates are both applicable and preferable, it is not always the case that the "ageing" process causes the maximum amount of mordant to be fixed upon the fibre. In such cases, other modes of getting rid of the acid are adopted, and other mordanting salts even are employed. The mordanting base may be fixed upon the fibre by using a weak alkaline bath of ammonia, chalk, sodium carbonate, etc., or by using such alkaline salts as not only remove the acid, but also produce insoluble compounds with the base, as sodium silicate, phosphate, arsenate, etc. This method is adopted by the calico-printer in the operation of "cleansing" or "dunging," which succeeds the "ageing" and precedes the dyeing operations.

When the mordanting body is applied in alkaline solutions—as stannic oxide, as stannate of soda—a slightly acid bath (sulphuric acid) is required for its precipitation upon the fibre. This is a method also frequently used by the calico-printer.

Still another method of fixing the mordant on textile fabrics is that of steaming, a process adopted for certain styles of work by the printer of cotton, wool, and silk materials.

In the calico styles referred to, a mixture of polygenetic colouring matter (as Alizarin) and metallic salt (as aluminium acetate) is printed upon the fabric, which is then dried and submitted to the action of steam in a closed box. During this steaming process, the metallic salt employed as mordant is decomposed, a greater or less proportion of its acid is driven off, and the remaining oxide or basic salt is fixed upon the fibre. Not only so, however, but at the high temperature employed, combination between the colouring matter and mordant takes place, coloured pigment is produced, and is at the same time firmly fixed upon the fibre.

The mordanting and dyeing operations are combined in an analogous manner when applying certain colouring matters to wool by dyeing, since this fibre possesses the property of decomposing acid solutions of colour-lakes, and even of attracting and mechanically fixing the latter when undissolved, if sufficiently finely divided.

Colour Acids and Colour Bases.—In the above cases, where polygenetic colouring matters are employed, the actual mordants fixed on the textile fibre have more or less a basic character; as already stated, they are metallic oxides or basic metallic salts, and although these colouring matters are not really acids, but rather bodies of an alcoholic or phenolic nature, they possess so much of the acid character that they combine with these and other bases; it is not at all improbable indeed that the colouring matter and the mordant (when this is necessary) must always bear some such definite relationship towards each other. All polygenetic colouring matters known hitherto possess the acid character referred to.

It has been stated that cotton does not become permanently dyed when immersed in a hot solution of Indigo Extract or of Magenta. In so far as this latter colouring matter is a red-coloured body, although soluble, it may be considered analogous to the Alizarin-red pigment produced by the combination of alumina and Alizarin. The question arises, is it similarly constituted? is it produced by the combination of a basic body with one of an acid character ? Experiment answers yes, and shows it to be a chemical compound of a colourless base rosaniline with hydrochloric acid. It does seem, therefore, to have a constitution somewhat analogous to that of Alizarin-red, but of a reverse character. In Magenta, the colouring power resides in the basic part of the compound (rosaniline), whereas in the Alizarin-red it is to be found in the acid portion (Alizarin), although in each case the other constituent is equally necessary to the production of a coloured body. Such considerations lead one to distinguish colour-acids and colour-bases, and we may infer that if the former, as we have seen, require basic mordants, the latter will probably require acid mordants. Among the numerous monogenetic colouring matters there is an extensive class of colour-acids which differ considerably in chemical constitution from those which possess an alcoholic or phenolic character like Alizarin. They contain the atomic group (HSO_3), are analogous more or less to acid sulphites, and have been termed "sulphonic acids." To this class belong Indigo Extract, Croceïn Scarlet, etc. Some colouring matters, as Indigotin, may be regarded as of a neutral or indifferent character.

In endeavouring to fix Magenta upon cotton, the question arises, will the colourless rosaniline combine with any other acid than hydrochloric acid to form an insoluble red or otherwise coloured compound? Is it capable of forming a lake? If so, the next question is, is the requisite acid capable of being fixed upon cotton in such a manner that it can still combine with the rosaniline? Experiment shows that there are such acids, as tannic acid. If a solution of Magenta is mixed with a solution of tannic acid (either free or neutralised with an alkali), an insoluble red-coloured tannate of rosaniline will be precipitated. Cotton has a natural attraction for tannic acid, so that when once steeped in its solutions it is not readily removed by washing. In order to dye cotton, therefore, with Magenta, it suffices to immerse it for some time in a solution of tannic acid, and, after drying, to pass it into a solution of Magenta. The red tannate of rosaniline thus produced upon the fibre does not, however, possess the character of absolute insolubility, especially in alkaline and soapy liquids, so that the dye cannot be considered entirely satisfactory. But, just as it has been seen that certain alkali salts can be used for the better fixing of the basic mordants on cotton, by reason of their acid, so here certain metallic salts can be used to fix such acid mordants as tannic acid, but in this case by reason of their base.

In applying Magenta to cotton, for example, a dye much faster to boiling soap solutions is obtained if the tannic acid-prepared cotton is passed into a solution of an antimony or tin salt—such as tartar emetic, or stannic chloride—previous to its immersion in the solution of Magenta. By this means the tannic acid is fixed upon the cotton in a very insoluble form, as tannate of antimony or tin.

Acid mordants, which act in the same manner as tannic acid, and fix the basic colouring matters upon cotton, are not numerous, but oleïc acid and other fatty acids may be mentioned as such. It is interesting to note that colouring matters of an acid character (as Alizarin) when fixed on cotton, may also behave as mordants towards basic colouring matters. Alizarin purples and Alizarin reds on cotton can be readily dyed with Methyl Violet, Magenta, etc. Hitherto all the acid mordants employed to fix any particular basic colouring matter on textile fabrics have produced only similar shades of colour. Basic colouring matters are hence all monogenetic.

Since, however, both oleïc and tannic acid can combine not only with organic colour-bases in the manner just described, but also with certain metallic oxides (inorganic bases), to produce insoluble compounds, they may be, and are, indeed, employed as fixing agents for the latter in the same way as the alkaline phosphates, arsenates, etc.

A usual method of dyeing cotton black, for example, is first to impregnate the cotton with a solution of tannic acid (decoction of sumach, etc.), and afterwards with a solution of a ferric salt (nitrate of iron).

The mordant (ferric oxide) is in this way fixed on the cotton by means of the tannic acid. Thus mordanted, the cotton is ready to be dyed in a decoction of logwood.

Another notable example of the same kind is afforded by the method employed in dyeing Turkey-red. Here the cotton is first impregnated with oleïc acid, or other oil compound of a similar character, and is afterwards immersed in a solution of an aluminium salt. The mordant alumina is fixed on the cotton by means of the oil compound, and yet it combines with the Alizarin in the subsequent dye-bath to produce the red pigment.

Apart from this preliminary precipitation and fixing of the basic or acid mordant on the cotton previous to the application of a colour-acid or colour-base, the fixing of all colouring matters upon cotton seems to depend largely on their capability of forming insoluble precipitates or lakes.

Colouring matters, like Indigo Extract, Croceïn Scarlet, etc., which do not form any sufficiently insoluble compound with bases, are not suitable for dyeing cotton.

In dyeing with Indigo and Safflower, the colouring matters are themselves readily precipitated from their solutions, either by oxidising or acid influences.

With Turmeric, and some few other dye-stuffs, precipitation is not necessary, since cotton is dyed with these by merely steeping it in their decoctions, and the case seems to be analogous to the dyeing of wool with Magenta.

INDEX.

Acid, Action of, on Cotton, 14
—, —— ——, —— Jute, 27
—, —— ——, —— Silk, 66, 67
—, —— ——, —— Wool, 38
—, Oxalic (see Oxalic Acid)
—— Mordants, 154
—, Nitric (see Nitric)
—— Salts in Water, 129
—, Sulphuric (see Sulphuric)
Alkaline Carbonates in Water, 129
—— Solutions, Scouring Wool with, 98, 99
Alkalis, Action of, on Cotton, 16-18
—, —— ——, —— Silk, 67, 68
—, —— ——, —— Wool, 39
Ammonia, Caustic, 17, 18
Amyloid, 14
Aqua Regia for Bleaching Silk, 121
Atlas Silk, 53
Barium Binoxide for Bleaching Tussur Silk, 123
Baur on Chemical Retting, 22, 23
Bleaching Action of Sulphur Dioxide, 117
—— Calico, 74
—: Chemicking, 73, 74
—, Cotton, 12, 71-87
—— by Electrolysis, 87
—— Flax, 26
—— Jute, 27
—: Ley Boil, 72
—— Linen, 89-92
—— ——, Chemistry of, 91, 92
—— ——, Processes in, 88
—— ——, Reeling Machine for, 89
—— ——, Rubbing in, 91
—, Liquid, 117
—, ——, Hydrogen Peroxide for, 118
—: Madder-bleach, 74
—, Mather and Platt Electrolyser for, 87
—— Oëttel Electrolyser for, 87
—— Powder, 18, 83
—— Raw Cotton, 71
—, Silk, 121, 122
—, Singeing before, 74-78
—: Souring, 73, 74
—— Tussur Silk, Motay's Method of, 123

Bleaching Warps, 71
—— Wool, 115-118
—— Yarn, 116
—— ——, Sulphur for, 116
Bluing Machine, 74
Boiling-off Silk, 119, 120, 121
Bolley on Silk Composition, 64
"Breaking" Flax, 23
Calcareous Impurities in Water, 125-128
Calico, Bleaching, 74
Cashmere, 33
Caustic Ammonia, Action of, on Cotton, 17, 18
—— Soda, Action of, on Cotton, 16
—— —— for Purifying Water, 134
Cellulose, 12
—, Analysis of, 13
—, Hydro, 14
—, Impurities in, 12
—, Nitro, 15
—, Oxy, 16
—, ——, Witz on, 18
Chandelon on Wool Scouring, 102
Chemical Retting, 22
Chemicking, 73, 74, 83
Chevreul's Wool Analysis, 41
China Grass, 27, 28
Chlorine, Action of, on Cotton, 18
—, —— ——, —— Wool, 39, 40
Clark's Process of Purifying Water, 132, 133
—— Scale of Water Hardness, 126
Cloth Scouring, 111-113
Cocoon, Silk, 50, 51
Cold-water Retting, 21
Collodion, 15
Colour Acids for Mordanting, 152
—— Bases for Mordanting, 152
Colouring Matters, Action of, on Cotton, 69
—— ——, —— ——, —— Silk, 69
—— ——, —— ——, —— Wool, 40
—— —— used in Dyeing, 145
—— Principles in Dyeing, 147, 148
Conditioning Apparatus, Silk, 60
Corron's Machine for Shaking Out Silk, 55
Cotton, 9-19
—, Action of Acids on, 14
—, —— —— Alkalies on, 16-18

INDEX.

Cotton, Action of Caustic Ammonia on, 17, 18
—, — — Soda on, 16
—, — — Chlorine on, 18
—, — — Colouring Matters on, 19
—, — — Frost on, 13, 14
—, — — Hypochlorites on, 18
—, — — Lime on, 18
—, — — Metallic Salts on, 18
—, — — Mildew on, 13
—, — — Nitric Acid on, 14, 15
—, — — Oxalic Acid on, 15
—, — — Sulphuric Acid on, 14
—, Bleaching, 12, 71-87 (see also Bleaching)
—, — Powder for, 18
—, Bluing Machine for, 74
—, Chemical Composition of, 12
— Cloth, Bleaching, 74
—, Dyeing Black, 154
—: "Extracting," 14, 19
— Fibres, 11
— —, Dead, 11
—, Fixing Magenta on, 153, 154
—, Impurities in, 12
—, Mercerised, 16, 17
—, Mercer's Process for, 16
—, Mordanting, 19, 150, 151
—, Physical Structure of, 11
— Plant, 9, 10
— —, Varieties of, 10
—, Raw, Bleaching, 71
Crabbing Machine, Treble, 113
Cramer's Formula for Fibroïn, 65
Cross and Bevan on Substance of Jute, 26
Crum Brown's Analysis of Purified Water, 143
Dead Fibres, 11
Degumming Silk, 119, 120
Dew Retting, 22
Dextrin, 14
Disease, Wool-sorter's, 34
Duseigneur on Silk, 49
Dye-houses, Purifying Water from, 137, 140-144
Dyeing, 145-155
—, Colouring Matters used in, 145
—, — Principles in, 147, 148
— Cotton Black, 154
—, Materials for, 145
—, Pigments used in, 147, 148
— with Turmeric, 155
Ecru Silk, 122, 123
Electrolysers for Bleaching, 87
Electrolysis, Bleaching by, 87
Eria Silk, 53
Expander, Mycock Five-bar, 85, 86
"Extracting," 19
Fat, Wool, 43
"Felting" of Wool, 31
Ferruginous Impurities in Water, 128, 129
Fibres, Cotton, 11

Fibres, Dead, 11
—, Flax, Retting, 20-23
Fibroïn, 63-65
—, Cramer's Formula for, 65
Fischer's Furnace for Yolk-ash, 102
Flax, 20-26
—, Bleaching, 26
—, "Breaking," 23
—, Chemical Action on, 25, 26
—, — Composition of, 25
— Fibres, "Retting," 20-23
—: "Grassing," 22
—: Hackling, 23, 24
— Plant, 20
—, Physical Structure of, 24, 25
—, Properties of, 24, 25
—: "Retting," 20-23
—: —, Chemical, 22
—: —, Chemistry of, 23
—: —, Cold-water, 21
—: —, Dew, 22
—: —, Stagnant-water, 21, 22
—: —, Warm-water, 22
—: Rippling, 20
—: Scutching, 23
—: "Spreading," 22
Flax-line, 24
Fleece Wool, 36
Foreign Wool, 33
Frost, Action of, on Cotton, 13, 14
Gaillet and Huet's Water Softening Process, 136
German Wool, 36, 37
Glossing Silk, 55
Glucose, 14
Gossypium Arboreum, 10
— *Barbadense*, 10
— *Herbaceum*, 10
— *Hirsutum*, 10
— *Peruvianum*, 10
— *Religiosum*, 10
Grass, China, 27, 28
"Grassing" Flax, 22
"Grey-sour," 80
"Grey-washing," 78
Gun-cotton, 15
—, Kuhlmann on, 15
Hackling Flax, 23
Havres on Wool-scouring, 100, 102
Havrez' Method of Utilising Yolk, 45, 46
Hydro-cellulose, 14
Hydrochloric Acid, Action of, on Silk, 67
Hydrogen Peroxide for Liquid Bleaching, 118
Hygroscopicity of Wool, 34
Hypochlorites, Action of, on Cotton, 18
—, — —, — Wool, 39, 40
Injector Kier, 81
Iron, Testing for, in Water, 128
Jute, 26, 27
—, Action of Acids on, 27
—, Bleaching, 27
—, Cross and Bevan on Substance of 26

"Kemps," 32, 33
Keratin, 36
Kier, Injector, 81
——, Mather and Platt, 79
——, Vacuum, 81
Kolb's Experiments on Retting Flax, 23
Kuhlmann on Gun-cotton, 15
Ley, or Lye-boil, 72, 80, 81
Lime, Action of, on Cotton, 18
——, —— ——, —— Wool, 39
—— for Purifying Water, 132, 133
Lime-boil, 79
Lime-sour, 80
Linen Bleaching, 89-92
—— ——, Chemistry of, 91, 92
—— ——, Reeling Machine for, 89
Liquid Bleaching, 117
Lustre of Wool, 35
Lustring Silk, Machine for, 58
—— without Tension, 95
Lye Boil, 72, 80, 81
McNaught's Wool-scouring Machine, 103, 105
Madder-bleach, 74
——, Time required for, 86
Magenta, Fixing, on Cotton, 153, 154
Magma Process of Wool Scouring.
Magnameries, 47 106
Magnesian Impurities in Water, 125-128
Marcker and Schulz's Analyses, 42, 44
Market-bleach, 86
Mather and Platt's Electrolyser for Bleaching, 87
—— —— —— Kier, 79
—— —— —— Singeing Machines, 76-78
Maumene and Rogelet's Analysis of Yolk-ash, 44
Mercerised Cotton, 16, 17
Mercerising, 93-95
Mercer's Process for Cotton, 16
Merino Wool Fleece, 33
Metallic Salts, Action of, on Cotton, 18
—— ——, —— ——, —— Silk, 68, 69
—— ——, —— ——, —— Wool, 40
Mildew, Action of, on Cotton, 13
Mohair, 33
Mordanting, 149-155
——, Colour Acids for, 152
——, —— Bases for, 152
—— Cotton, 19, 150, 151
—— Textile Fabrics, 152
—— Woollen Fibre, 149, 150
Mordants, Acid, 154
Motay's Method of Bleaching Tussur Silk, 123
Moth, Silk, 47
Muga Silk, 53
Mulberry Silk, 54
Mulder's Analyses of Silk, 63, 64
Mullings' Method of Wool Scouring, 106, 107

Mycock Five-bar Expander, 85, 86
—— Scutcher, 84
Nitric Acid, Action of, on Cotton, 14, 15
—— ——, —— ——, —— Silk, 66
—— ——, —— ——, —— Wool, 38
Nitro-cellulose, 15
Oëttel Electrolyser for Bleaching, 87
Organzine, 51, 52
"Over-retting," 22
Oxalic Acid, Action of, on Cotton, 15
Oxy-cellulose, 16
——, Witz on, 18
Parchment, Vegetable, 14
Permanganic Acid, Action of, on Silk, 67
Peroxide of Hydrogen, 118
Perspiration, Wool, 43-45
Pigments used in Dyeing, 147, 148
Plant, Cotton, 9, 10
Porter-Clark Process of Softening Water, 135, 136
Powder, Bleaching, 18, 83
Pyroxylin, 15
——, Soluble, 15
Reagents, Influence of, on Silk, 66
Red-bleach, Turkey, 86, 87
Reeled-silk, 52
Reeling Machine, 89
Retting, Chemical, 22
——, ——, Baur on, 22, 23
——, Chemistry of, 23
——, Cold-water, 21
——, Dew, 22
—— Flax, 20-23
——, Kolb's Experiments on, 23
——, Stagnant-water, 21, 22
——, Warm-water, 22
——, ——, Schenck on, 22
Rhea, 27, 28
Rippling Flax, 20
Rubbing in Bleaching Linen Cloth, 91
Sanderson's Dye-works, Water Purification at, 141-143
Schenck on Warm-water Retting, 22
Schwamborn's Dye-works, Purification of Water at, 144
Scrooping of Silk, 54, 55
Scouring, 46
—— Agents for Wool, 97
—— Bath, 98
—— Cloth (see Cloth Scouring)
—— Loose Wool, 98-107
—— Silk, 119-123
—— Substances, 98
—— Union Material, 113-115
—— Wool (see Wool)
—— Yarn (see Yarn)
Scutcher, Mycock, 84
Scutching Flax, 23
Sericin, 65, 66
Shaking Out Silk, 55
Silk, 47-70

INDEX. 159

Silk, Action of Acids on, 66, 67
——, —— —— Alkalis on, 67, 68
——, —— —— Colouring Matters on, 69
——, —— —— Hydrochloric Acid on, 67
——, —— —— Metallic Salts on, 68, 69
——, —— —— Nitric Acid on, 66
——, —— —— Permanganic Acid on, 67
——, —— —— Water on, 66
——, —— —— Zinc Chloride on, 68
——, Aqua Regia for Bleaching, 121
——, Atlas, 53
——, Barium Binoxide for Bleaching Tussur, 123
——, Bleaching, 121, 122
——, Boiled-off, 119
——, Boiling-off, 120, 121
——, Bolley on Composition of, 64
——, Chemical Composition of, 62, 63
—— Cocoon, 50, 51
—— ——, Classes of, 51
——, Conditioning, 61, 62
——, —— Apparatus for, 60-62
——, Corron's Machine for Shaking out, 55
——, Culture of, 47
——, Degumming, 119, 120
——, Duseigneur on, 49
——, Dyed, Examination of, 69, 70
——, Ecru, 122, 123
——, Elasticity of, 59
——, Eria, 53
——, Glossing, 55
——, Influence of Reagents on, 66
——, Lustring, 58
——, —— Machine for, 58
——, Motay's Method of Bleaching Tussur, 123
—— Moth, Eggs of, 47
——, Mulberry Silk, 54
——, Muga, 53
——, Mulder's Analyses of Composition of, 63, 64
——, Origin of, 47
——, Physical Properties of, 54, 55
——, Raw, 51, 52
——, Reeled, 52
—— Scouring, 119-123
—— Scrooping, 54, 55
——, Sericin, 65, 66
——, Shaking Out, 55
——, Softening, 121
——, Solvent for, 69
——, Souple, 121, 122
——, Soupling, 122
——, Specific Gravity of, 58
—— Spinning, 47-50
——, Stoving, 121, 122
——, Stretching, 121
——, Stringing, 55
——, —— Machine for, 57, 58
——, Tenacity of, 59
——, Tussore, 53

Silk, Tussur, Bleachi 123
——, Waste, 52, 53
——, Wild, 53, 54
——, Winding, 51, 52
——, Yama-maï-, 53
Silk-glue, 65
Silkworms, Rearing, 47, 48
Singeing, 74-78
—— Machines, Mather and Platt's, 76-78
——, Washing after, 78
Sodium Carbonate as Scouring Agent, 98
Soluble Pyroxylin, 15
Solvent for Silk, 69
Souple Silk, 121, 122
Soupling Silk, 122
Souring, 73, 74
Spindler's Dye-works, 140, 141
"Spreading" Flax, 22
Stagnant-water Retting, 21, 22
Steaming Union Material, 115
Stretching Silk, 121
—— Yarn, Machine for, 108
Stringing Machine for Silk, 57, 58
—— Silk, 55
Steeping, 99-103
Stoving Silk, 121, 122
Sulphur Dioxide, Action of, on Wool, 38, 39
—— —— as Bleaching Agent, 116
—— in Wool, 37
Sulphuretted Hydrogen in Water, 130
Sulphuric Acid, Action of, on Cotton, 14
Tanks for Wool-steeping, 99-101
Thread, Linen, Bleaching, 88
Tram, or Weft-silk, 51, 52
Treble Crabbing Machine, 113
Turkey Red-bleach, 86, 87
Turmeric, Dyeing with, 155
"Turn-hanking," 91
Tussore Silk, 53
—— ——, Bleaching, 123
Union Material, Crabbing, 113
—— ——, Scouring, 113-115
—— ——, Steaming, 115
Urine as Wool Scouring Agent, 97, 98
Vacuum Kier, 81
Vegetable Parchment, 14
Vetillart on Flax Structure, 24, 25
Volatile Liquids used in Wool Scouring, 106, 107
Warm-water Retting, 22
Warps, Bleaching, 71
Washing, Final, before Bleaching, 83, 84
—— after Singeing, 78
—— Wool, 103
Wash-water Products of Raw Wool, 45
Waste Silk, 52, 53
Water, 124-144
——, Acid Salts in, 129
——, Action of, on Silk, 66

Water, Alkaline Carbonates in, 129
—, Calcareous Impurities in, 125-128
—, Caustic Soda for Purifying, 134
—, Chemical Purification of, 132
—, Clark's Process of Purifying, 132, 133
—, — Scale of Hardness of, 126
— from Dye-houses, Purifying 137, 140-144
—, Effluent, Crum Brown's Analysis of, 143
—, Ferruginous Impurities in, 128, 129
—, Hard, 125
—, Impurities in, 125-130
—, Iron in, 128
—, Lime for Purifying, 132, 133
—, Magnesium Impurities in, 125-128
—, Mechanical Purification of, 131
—, Natural Impurities in, 125
—, Organic Impurities in, 130
—, Porter-Clark Process of Softening, 135, 136
—, Purifying, 130, 131
—, — by Boiling, 131, 132
—, — with Caustic Soda, 134
—, — with Lime, 132, 133
—, —, at Sanderson's Dye-works, 141-143
—, —, — Schwamborn's Dye-works, 144
—, — — Spindler's Dye-works, 140, 141
—, Soft, 124
—, Softening, by Gaillet and Huet's Process, 136, 137
—, —, — Porter-Clark Process, 135, 136
—, Sulphuretted Hydrogen in, 130
Weft-silk or Tram, 51
White-sour, 83
Wild Silk, 53, 54
Winding Silk, 51, 52
Witz on Oxy-cellulose, 18
Wool, 29-46
—, Action of Acids on, 38
—, — — Alkalis on, 39
—, — — Chlorine on, 39, 40
—, — — Colouring Matters on, 40
—, — — Heat on, 37, 38
—, — — Hypochlorites 39, 40
—, — — of Lime on, 39
—, — — Metallic Salts on, 40
—, — — Nitric Acid on, 38
—, — — Sulphur Dioxide on, 38, 39
— Bleaching, 115-118

Wool Bleaching, Sulphur Dioxide Agent for, 116
—, Chemical Composition of, 36
—, Chevreul's Analysis of Raw, 41
—, Elasticity of, 35
— Fat, 43
—, "Felting" of, 31
—, Fleece, 36
— —, Merino, 33
—, Foreign, 33, 34
—, German, 36, 37
—, Hygroscopicity of, 34
—: "Kemps," 32, 33
—, Loose, Scouring, 98-107
—, Lustre of, 35
—, Marcker and Schulz's Analysis of Raw, 42
—, Physical Structure of, 30, 31
—, Raw, Chevreul's Analysis of, 41
—, —, Substances found in, 41, 42
—, —, Wash-water Products of, 45
—, —, Yolk in, 40
— Scouring, Chandelon on, 102
— —, Havrez on, 100, 102
— — Machine, McNaught's, 103, 105
— —, Magma Process of, 106
— —, Mullings' Method of, 106, 107
— — with Volatile Liquids, 106, 107
—, Sulphur in, 37
—, Value of, 36
—, Varieties of, 29, 30
—, Washing, 103
—, Yolk in Raw, 40
Woollen Cloth Scouring Machine, 112, 113
— Fibre, Mordanting, 149, 150
Wool-sorter's Disease, 34
Wool-steeping Tanks, 99-101
Yama-maï Silk, 53
Yarn, Bleaching, 116
—, —, Sulphur for, 116
—, Linen, Bleaching, 88, 89
— Scouring, 107-111
— — Machine, 110, 111
— Stretching, 108
— — Machine, 108
Yolk, Havrez's Method of Utilising, 45, 46
— in Raw Wool, 40
Yolk-ash, Analyses of, 44
—, Fischer's Furnace for Making, 102
Zinc Chloride, Action of, on Silk, 68